ARCTIC OCEAN ENGINEERING FOR THE 21ST CENTURY

BEN C. GERWICK, EDITOR

PROCEEDINGS OF THE

FIRST SPILHAUS SYMPOSIUM

PRESENTED BY THE

MARINE TECHNOLOGY SOCIETY

ISBN: 0-933957-00-9

This work relates to Department of Navy Grant N00014-84-G-0216 issued by the Office of Naval Research. The United States Government has a royalty-free license throughout the world in all copyrightable material contained herein.

Marine Technology Society
2000 Florida Ave., N.W.
Suite 500
Washington, D.C. 20009

Contents

PREFACE

Arctic Ocean Engineering for the 21st Century is the proceedings volume of the First Spilhaus Symposium held October 14–17, 1984 in Williamsburg, Virginia. Conceived of and presented by the Marine Technology Society, the Spilhaus Symposium Series takes its name and—we hope—its "practical visionary" spirit and energy from Dr. Athelstan Spilhaus, outstanding engineer, scientist, writer, inventor and humanist. The series is intended to provide a forum for innovative marine scientists and engineers to express freely their ideas, beliefs and plans for the future. It is hoped that such frank and forward-looking discussions will stimulate new approaches to the challenges of today and tomorrow in the area of marine technology.

Arctic Ocean Engineering is a particularly apt topic for the First Spilhaus Symposium because of the tremendous strategic importance of the region worldwide and because its remoteness and harsh environment demand long lead times as well as coordination and cooperation between international experts with diverse specialties in industry, academia and government.

The success of the Symposium, as reflected in the high quality of the participants, the spirited discussions which ensued, and the merits of this volume can be largely attributed to the many talents and tireless efforts of technical program chairman Ben C. Gerwick, Jr.

Although a plethora of Arctic Ocean Engineering conferences has addressed the current state-of-the-art and short-term Arctic development, the Spilhaus Symposium was different in method and focus. Based on the Gordon Conferences in Science, the meeting was loosely structured to provide participants an opportunity, in Ben Gerwick's words, "to allow imagination to range freely into the future."

The heart of the conference was half-day workshops which focused on the topics of Work Systems, Oil and Gas, Environmental Considerations, Transportation, Scientific Research, Minerals, Sea Ice Management and Materials. Experts in these areas gave brief technical presentations, then workshop leaders led discussions exploring research and development needs, priorities and constraints over the next twenty years.

This conference record reflects the organization of the symposium. It is divided into four parts: Arctic Perspectives, Workshop Reports, Closing Plenary Session, and Appendices. The first consists of technical presentations by invited speakers; the second summarizes the discussions and conclusions of each workshop; the third encompasses technical presentations, general discussion and closing remarks; and the Appendices include a list of attendees, a copy of the *Arctic Research and Policy Act of 1984*, background abstracts, and technical presentations voluntarily submitted by symposium participants.

It is our hope that this conference record preserves not only the technical content of the discussions, but also the excitement and enthusiasm at the First Spilhaus Symposium.

Mary Beth Hatem
Wendy E. Hudson
Technical Editors

ACKNOWLEDGMENTS

The Marine Technology Society gratefully acknowledges the generous support of The Max and Victoria Dreyfus Foundation, Inc.; the National Oceanic and Atmospheric Administration; the U.S. Department of Energy; the U.S. Geological Survey; and, the U.S. Department of Navy.

The Society also wishes to extend its gratitude to the dedicated and enthusiastic efforts of the Symposium's Program Committee:

GENERAL CHAIRMAN:

Dr. Harris B. Stewart, Jr.
Old Dominion University

TECHNICAL PROGRAM CHAIRMAN:

Mr. Ben C. Gerwick, Jr.
University of California at Berkeley

TECHNICAL PROGRAM COMMITTEE:

Lt. Cdr. Lawson W. Brigham
Ice Operations Division, U.S. Coast Guard

Dr. Ira Dyer
Massachusetts Institute of Technology

Mr. George Gryc
U.S. Geological Survey

Mr. S.M. Maxwell
Gulf Canada Resources, Inc.

Mr. M. Oshima
Mitsui Engineering and Shipbuilding Co., Ltd.

Dr. Ernest Pallister
Pallister Resource Management, Ltd.

Dr. William Sackinger
University of Alaska

Dr. George Vance
Mobile Research and Development Corp.

MTS MANAGEMENT COMMITTEE:

Mr. Hank Porterfield, Chairman
BEK Associates

Mr. Steven N. Anastasion
NACOA

Mr. John E. Flipse
Texas A&M University Research Foundation

ADVISORY COMMITTEE:

Mr. Gilbert L. Maton
Tracor-Jitco, Inc.
Mr. Richard M. Shamp
Engineering Services Associates, Inc.

PART I:

INTRODUCTION AND BACKGROUND

THE SPILHAUS SYMPOSIA CONCEPT

Ben C. Gerwick, Jr.
University of California at Berkeley

These Spilhaus symposia are an idea engendered by Dr. Athelstan Spilhaus and Gilbert Maton, the President of the Society. They envisioned a series of symposia which would address the developments in Ocean Engineering in the 21st century. This is the first of these symposia and, as you know, is directed to the Arctic Ocean. Their concept was to pattern these after the Gordon conferences which have been so successful in science. As I think many of you are aware, the Gordon conferences are usually limited to an attendance of 80 people: eminent scientists in that particular branch or discipline. Just as we are doing here, the Gordon conferences are purposefully run on an informal, unstructured basis to a large degree, so that they will encourage the interchange between experts of ideas that are in the process of development.

Now, all of us have previously attended conferences in which we have had to go through the process of preparing formal papers that have been reviewed, documented, and properly referenced. This is not at all the concept here. The concept intended here, as in the Gordon conferences, is to develop spontaneous and long-range ideas, to move ahead. That is why we are encouraging as much informal contact as possible, leaving time at lunches and in the evenings for each of you to exchange your ideas with your colleagues.

As we approach Arctic and Ocean Engineering in the 21st century, we recognize that there are a plethora of conferences on this subject. We've had two OTC specialty sessions on Arctic engineering, and ASME's conference on ice mechanics. Next May we are having the ASCE Specialty Conference on the Arctic. The Department of Energy has just announced its conference just a few weeks from now in Morgantown, West Virginia. However, all of these are aimed at a short-range state of the art and the current needs. This conference does differ, then. We are addressing the long-range, and we are trying to project our imaginations into the future.

Now, you've all heard it said that we are too optimistic when we address next year. We are too sure of all the things we can accomplish within a one-year period. Conversely, we are very bold when we address something that may take place in 100 years. One hundred years from now, we will have no particular responsibility, so we can let our imaginations run free. A more difficult range to address is that of the 15- to 20-year period for which we do have the responsibility of starting the developments now.

Last week I was in another part of the world which shares many of the environmental aspects of the Arctic. I was in Saudi Arabia. Saudi Arabia, like the Arctic, is a vast, almost featureless area. It is subject to extremes of temperature. It is a remote part of the world, so that logistics are extremely difficult. It has a small, indigenous population, so that most of the technological development has to come from outside. When one flies over Saudi Arabia, it almost looks like flying over the Arctic at wintertime, except that the featurless plain is slightly yellowish instead of white.

In 1950, when I first went to the Middle East, the emphasis was solely on trying to find oil. Shortly thereafter, as oil began to flow in large quantities, the emphasis shifted to short-term extraction. I can recall discussing with one of the large groups of oil companies the building of certain facilities, for which I was trying to emphasize durability for a ten-year period. I was told to forget it; that if they couldn't maximize their extraction within three years, the project wouldn't go ahead.

I think we can see a certain parallel between the development in Saudi Arabia in the '50s, '60s, and '70s and what is starting now in the Arctic just 20 years behind. It is just offset in the time scale.

Now, returning to Arabia for a moment, we had oil extraction in the '50s, the development of large, permanent shipping facilities and permanent processing plants in the '70s, and then at the end of that period, the idea of an infrastructure. Who could ever have imagined that there would be a city, in this case Jubail, with a permanent population of 60,000 people living in modern housing with five major world-scale industrial plants, with airports and hospitals and schools, all having sprung to life in the most inhospitable place in the world?

Now, I personally have a great deal of difficulty envisioning large cities in the Arctic of that scale. Maybe my imagination is not sufficient. And yet we already know that Prudhoe is one of the major cities in Alaska. The Soviets are reportedly developing one major new city on the Arctic ocean each year. So let's not limit ourselves as we think ahead 20 years.

Engineering, which is our main focus here, cannot exist in a vacuum. We are constrained by all the limitations of our social, ecological, and political environment, and that's why we've set up this program and asked some eminent people to give us an overview of these various factors which will so deeply influence what we can do in a technological sense. We also need to be involved with Arctic science, because the data upon which we base our work is scientific, and in turn, science needs the support of engineering: it needs platforms and instrumentation and vehicles and communication.

The Arctic is characterized by extremely long lead times. The *Manhattan* made its trip in 1970, and yet we still have not actually produced oil from offshore in the Arctic. The year 2000 is only 15 years away.

Let's look for a moment at the objectives of this conference. We are trying to identify trends, opportunities, we are trying to look at the interaction between disciplines and the interaction between nations. I have only one disappointment in this conference, and that is that we were not able to arrange for participation here by engineers from the Soviet Union; perhaps this was the wrong year and the wrong time for that. But nevertheless, as we look at the long-range future, perhaps it is the Arctic with its formidable challenges and demands extending beyond national boundaries, which will give us the opportunity for working together in the future and for finding means of truly international cooperation.

Now, in carrying out this symposium, we start this morning with broad, overview papers, and I think you will be as pleased as I am with the caliber and the breadth of view that will be expressed. We will then be having workshops, which you have all seen on the program, devoted to Arctic work systems, Arctic oil and gas, environmental considerations, transportation, scientific research, Arctic minerals, sea ice management, and Arctic materials.

During these workshops, we want to encourage participation from all those present. The reason for having workshops is to divide into small enough groups that people can really speak out, can ask questions, and present ideas. This will not be a successful conference if it only consists of a few presentations by those whom we have asked to give their ideas ahead of time. We must get some spontaneity into this program.

And with that in mind, may I also encourage individual exchange during the coffee breaks, lunches and dinners? At the end of this meeting on Wednesday, we will try to come together and identify those particular aspects that are new and promising and to reach some sort of consensus as to where we should be going in our technological development of the Arctic.

What product do we hope for at the end? Yesterday as I drove down I-95 and turned off onto 295, I saw in the mirror coming behind me a car driving about 90 miles an hour. He couldn't decide which side he was going to pass me on, and he finally passed me on the left shoulder but almost immediately realized, as he jammed on his brakes, that he had taken the wrong turn. I think maybe there is a little lesson there; we don't want to run full speed into the Arctic in the wrong direction.

The Arctic is a challenge. Toynbee has said that societies come to their highest stage of development as they face and overcome challenges. I hope this conference will contribute to that by developing some meaningful contributions to Arctic Ocean Engineering in the 21st century.

Athelstan Spilhaus: An Introduction

Gilbert L. Maton
President, Marine Technology Society

It is always a privilege to introduce Dr. Athelstan Spilhaus. You will notice that this is billed as the first Spilhaus Symposium. The concept was originally generated by Spilly, but he did not expect us to name the Symposium for him. Spilly is a founder of the Marine Technology Society and if, today in the United States, we have anyone comparable to Ben Franklin, I think it is Athelstan Spilhaus, a Renaissance Man. He is an author of numerous books and articles—many of you might remember him best for ''Turn to the Sea'' and his weekly cartoon series, ''Our New Age,'' that inspired thousands of youngsters during the fifties and sixties. He is also a prodigious inventor, best known for the bathythermograph for measuring temperature versus water depth, and the Spilhaus Space Clock.

Spilly was born in Capetown, South Africa. He wouldn't allow, though, how many years ago that was. He did his doctoral work at the University of Capetown and spent some time at MIT doing additional work. Spilly founded the Department of Meteorology and Oceanography at New York University where he also served as Chairman from 1938–47. He went on for nearly 20 years as Dean of the Institute for Technology at the University of Minnesota where he retired in 1966.

I use the term ''retired'' loosely though. Spilly has been anything but that since 1966. He has been an advisor to three presidents and a member of numerous distinguished international committees and panels. Notably, he was appointed by President Johnson in 1966 to serve as a member of the National Science Board and, in addition, served as a member of the House Advisory Panel of the Committee on Science and Astronautics. He has numerous professional affiliations among which include the presidency of the Franklin Institute of Philadelphia and the American Association for the Advancement of Science. He has been honored for his military service and has been named a fellow of many professional societies, including the American Geophysical Union and the Marine Technology Society.

The list goes on and on, but most importantly, he is a valued friend and it is with great pleasure that I introduce today Dr. Athelstan Spilhaus, originator of the Spilhaus Symposium concept.

The Challenge Ahead: Arctic Science and Engineering

Athelstan Spilhaus
Founding Member, Marine Technology Society

I chose to go to sea when I was 16 on one of the ships of opportunity of those days—a 5,000-ton freighter plying from Capetown to Durban, Antanarive, Columbo, Calcutta, and Rangoon. With no air conditioning, it was hot as hell going across the equator. I don't know if anybody even remembers the ventilators they had on old ships. They were huge venturies, great things that you turned into the wind so that the wind blew in through a funnel device and sucked the hot air out of the engine room and out of the other rooms below.

I was sixth engineer. I think prior to my arrival they had gotten along very well with five. One of my jobs was to turn the ventilators into the wind. I learned about venturies from that, and it stood me in good stead. After graduating from the University of Capetown, I went to work in Germany in 1931, for the then famous Junkers Flugzeug Werke. Junkers was, I think, the leader in building the best commercial aircraft of that day. They made corrugated metal planes, and they were wonderful planes.

I was given a very responsible job on what then was a big transport plane, to design the first on-board aircraft toilet—a boon for mankind and his wellbeing. Of course, the main problem was how to draw out the noxious gases from the small compartment. So employing my engineering knowledge of how we sucked the gases out of the engine room at sea, yet knowing nothing about the scientific underpinnings, the V^2 law, or that the speed of an airplane is much greater than that of a cargo vessel, I built a great big venturi of which I was immensely proud.

On the test flight, we had onboard one of the big, fat company "direktors," with a gold watch chain across his belly. We had sausages and beer and black bread, and after awhile, to my great pleasure, I saw that the Direktor was going to test my contribution. He went into the toilet and, after an inordinately long time, there were noises indicating that he couldn't get the door open. The suction was so great it held the door in place, the result of a slight overdesign on my part. So we pried the door open, air rushed in from the cabin and he was sucked firmly to the seat. We had to slide him off. Almost immediately after that, I left Junkers. I decided I'd better go to MIT to learn about the scientific underpinnings of aerodynamics which are essential even in the design of in-flight toilets.

Perhaps there is a lesson in this true experience. We are here for engineering, but all good engineers are people who do the difficult job of blending the underpinnings of science with the doing and designing of things for the wellbeing of people.

I am most honored that all of you came here, a distinguished gathering. I am grateful to you, Gil, for the immense amount of trouble that you and all of the MTS people have taken to put this first conference together. I hope it will the forerunner of many more. It is up to us to make it successful.

Putting the logistics of a conference together is one thing, but the success of it depends on its intellectual integrity, the quality of the people who are invited to take part and the structure of the material to be discussed. For that task we were fortunate to get Ben Gerwick, a true engineer. An engineer has to look not only at making things work, but also doing things for society. He has to worry about societal concerns. He has to confront some good objections, political and environmental, and many stupid ones, political and environmental. But he nevertheless has to overcome them. He has to tangle with financial difficulties, and financial difficulties are not a small part of the difficulty of entrepreneurship in this day and age.

Professor Gerwick, in my opinion, is the embodiment of that wonderful word, "engineer." Engineers are also dreamers, and that's what we are here to do, but they are dreamers who can make dreams come true and I've seen many dreams come true. I recall in 1954 we had the first conference in connection with the International Geophysical Year in Rome where we proposed to put up a satellite for geophysical purposes, a tiny little basketball-sized object. There were some scientists there who said, "Well, we just can't put any electronics into this thing because in space the batteries will boil." I got very irritated at that, not because I knew a darn thing about batteries but because I had complete faith in engineers. I said, "Well, darn it, we'll get some engineers who will make batteries that won't boil," and we did. This is, I think, the kind of "can-do" attitude that we'd like to see at this conference.

There are going to be many challenges in the opening up of the Arctic, but let's regard them as challenges and find ways to overcome them and not see them as reasons for not doing anything. I think that what we are looking for really is a way to inhabit the Arctic, as has been done in Prudhoe Bay and by the Soviets in their Arctic. We want to go to the Arctic to stay. This is the kind of colonization which will best utilize this magnificent part of the world that is hardly scratched.

I am honored to have representatives of most of the countries bordering the Arctic at this symposium. They are the ones that should be here, starting with Finland and going *cum sole* around the pole, we have Norway, France, Britain, Canada, and the United States. A notable absence here is Denmark, with the great island of Greenland. I wish they were here. I am, as is Ben, sorry that our Russian colleagues are not here. We could have learned a lot from their operations in the Arctic. When we approach the problems of the Arctic, we do need to approach them on a cooperative basis. What can we as a group of nations with interests bordering on the Arctic do about it? It's a big enough job so that we have to do it together. Its rewards are so promising that there's plenty to share. I thank those who have come a long distance for their presence here.

Ben mentioned the folly of going full speed in the wrong direction. That's perfectly true. However, I will tell another story about traveling. Once I was driving a car and came to a level railroad crossing with gates to make traffic stop when the train was expected. I noticed the gate was down on my side, so I stopped the car. But the gate was up on the other side. I couldn't quite understand that. Why was the gate down on one side and up on the other? I wondered if I should just raise the gate, take a risk and go through, or if I should wait for the train. There didn't seem to be any train. Finally a man appeared who was obviously an official because he was wearing a cap. So I asked him, "How come you've got one gate down and one gate up?" He said, "Oh, I'm *half* expecting a train." Sometimes I think we don't do something we can do and for which we have the tools because we are half expecting disaster and not half working to overcome it.

Of course, this means taking risks. I think some of you may remember the rather silly period of Ralph Nader, the advocate of zero risk, who did not realize that zero, although marvelously useful in mathematics, does not exist in nature. In the Arctic, in attacking the Arctic, there are going to be risks—risks to life, financial risks, risks of capital investment, and risk of personal reputation. But the benefits are worth taking those risks. Nader approached the important problem of environment from the point of view of zero risk. Every good engineer knows part of good engineering is consideration for the environment. But Nader took a completely negative approach. Let us take a positive approach.

Recently I've had two great conversations; one was with the engineers that are doing the job of keeping the sea out of Holland. I always think this is one of the grandest engineering achievements having to do with the sea. Another new, daring and admirable project that the Japanese have undertaken is the building of the tunnel between Honshu and Hokkaido under the sea. These are the kinds of grand engineering feats that I think I dream of us doing, and the kinds of things we will have to tackle in the Arctic.

To the scientists I would say that sometimes scientists forget that, though their work is very important in itself, if it were not for engineers drawing out the good science and putting it to use for public wellbeing, the scientists wouldn't get the funds. The engineers sometimes forget, (as I did in the case of the airborne toilet) that they have to know some of the underpinnings of science. Certainly science and engineering are one.

I look at the program finally and I see a lot of "perspectives." Perspective comes from the root meaning "seeing through." I hope the perspectives we will hear about from the distinguished speakers this morning will help us see through to the future—visionary, positive perspectives of what we can do if we have the will to do it.

I see a lot of things that are called "policy." Policies in the absence of action are about as useless as that fifth appendage that they refer to on a cow. Policy, if it's formed in a vacuum (as it often is when it precedes experience), inhibits entrepreneurship in new activities.

I prefer, in a planning session of this nature, to speak of goals rather than policies. I would like to know what the Canadian goal is, what the American goal is, what the Norwegian goal is, what the Soviet goal is with respect to the Arctic. I can guess that all those goals are the same. I am sure that the Soviet goal, for example, is to move out into the Arctic, to obtain resources from the Arctic, and to use the Arctic as best they may, and that is what I see as our goal in the U.S. and I am sure what others see as their goal.

I think we should also guard against a repetition of that endless farce of the U.N. law of the sea. I say endless farce not because we don't badly need a law of the sea, but because it is being discussed in the wrong forum. The U.N. is the wrong forum because there are many, many nations that have a vote on the law of the sea that have no possible interest in the sea and simply use their bargaining chip, their vote, to further other political aims. It is very important that, when we approach this problem of the Arctic, we do not widen the international participation to those who have no possible interest in the Arctic, that we keep it to those nations that surround the Arctic and have a real interest in it. That is the way true law develops. As activity increases in an area, conflicts are bound to follow. These conflicts can be resolved by bilateral and multilateral agreements that will lead to the development of a body of law.

There will be problems on many sides due to military considerations. The military wields a two-edged sword—sword on one side, plowshare on the other. The military can be very inhibited because of secrecy, and our Navy and Air Force have sometimes been guilty of overdoing secrecy. In about 1951, the U.S. Air Force was making what we called Ptarmigal flights from Eilsen Air Force Base in Alaska, routinely, daily, over the North Pole, ostensibly for weather observations, but really to pick up debris because our intelligence knew that at about that time the Russians were going to explode a nuclear device or bomb. I flew on some of these flights. On one of these, my pilot, Joe Fletcher, spied the ice island later called Fletcher Island.

The Air Force immediately wanted to slap secrecy on this island. Fortunately, I was chairman of the scientific advisory committee to the Secretary of Defense and, at a meeting, I argued, ''Gentlemen, you can't place T-3, Fletcher's Island, under secrecy, because it won't fit in a three combination file cabinet.'' We managed to break it out.

These kind of things can happen, but let us not allow them to overcome us and use them as excuses for not doing anything. Let's just recognize they are going to happen and we will overcome them as they come along. The military interests use everything that civilian life uses. Any invention can be used for warlike or defense purposes or for peace-like purposes for the benefit of people. Again, because something is used for the military doesn't make it bad. One ocean engineering achievement that the United States did which was quite remarkable was carried out by the CIA when it raised that piece of a submarine from very, very deep in the Pacific Ocean, a great engineering achievement. The reasons they were able to succeed was that they had a purpose and the will. What we need here is purpose and will to open up the Arctic.

I very much look forward to listening to these discussions. I think you will uncover many challenges, but the tools are on the shelf to meet these challenges. Let's use them and go northwards to stay.

PART II:

ARCTIC PERSPECTIVES

The Arctic World—An Exclusive Frontier

Louis Rey
Comité Arctique International

THE ARCTIC OCEAN: A RECENT PHYSIOGRAPHICAL UNIT

Some 250,000,000 years ago, the Arctic regions were still unstructured. It was only towards the end of the Mesozoic period that continental plates, pushed northwards by the ridge activities, almost completely enclosed a vast section of the global sea originally surrounding the Pangea. Then, following the separation of North America from Eurasia, the Labrador Sea opened approximately 50,000,000 years B.P. (Before Present) and a deep Eurasian basin appeared in the Arctic Ocean—when the Lomonosov ridge became severed from Siberia.

The climate, which had been somewhat warm in the early Cenozoic became, during the mid-Cenozoic, a temperate boreal type climate with warm summers and cold winters. At the same time, in Antarctica, a mighty ice shelf spread over the whole Continent. In parallel, and as a function of the new distribution of land and sea masses in the Northern Hemisphere, an increased meridional flow developed within the global circulation.

This situation remained fairly stable for another 20 million years until, at the onset of the Pliocene, 7-5 million years ago, a steady decrease in temperature marked the beginning of the glacial age. Due to the increased meridionality of the air flow, the advection of warm, moist Atlantic air over cold, barren land masses resulted in the development of massive icecaps and high-latitude glaciers as early as 4.5 million years B.P. Concurrently the Antarctic ice sheet expanded once again and the overall result was a substantial eustatic drop of the mean level of the oceans.

The Polar Sea was then quite isolated and as yet had no communication with the Pacific and only a very limited exchange with the North Atlantic (the Norwegian-Greenland Sea being wide open only around 1,500,000 B.P.). Following paleontological evidence (calcareous nannofossils found in sedimentary cores from the crest and flank provinces of the Alpha Cordillera), Y. Herman describes the Arctic Ocean 5,000,000 years ago as a cold water expanse, displaying intense vertical mixing and high biological productivity due to the massive upwelling of nutrients and the excellent oxygenation of the entire water column. Sand-grain deposits from circumarctic glaciers indicate, however, substantial ice-rafting mechanisms and, therefore, a regular ice cover. Nevertheless, though ice-formation was quite considerable in winter time, there is absolute evidence that no perennial ice covered the Arctic Ocean until 3,000,000 years B.P.

Then, resulting from one—or several—of the still unknown mechanisms responsible for world climatic changes (variation of the earth's orbital position—changes in solar activity, atmospheric transparency, the earth's internal heat economy. . .) a sharp drop in temperature occurred approximately 3,000,000 B.P. At that time, the ocean circulation was also subjected to changes as a consequence of the elevation of the Isthmus of Panama and the opening of the Bering Strait (about 3,500,000 B.P.) which resulted in a new water transport into northern latitudes through, in particular, a more vigorous Gulf Stream.

For Y. Herman, the Polar Sea then became the site of a strong salinity-density stratification reinforced in summer by the spreading—over the underlying salty Atlantic waters—of a vast quantity of fresh water coming from the Arctic Rivers and the melting icecaps and, to some extent, due to the input of low-salinity Pacific waters. This initial ''conditioning'' was over-emphasized by the progressive melting of continental icecaps as a result of deglaciation. The obvious consequences of this stratification were two-fold: It prevented vertical mixing and, thus, seriously altered the biological production in the oxygenated nutrient-deprived surface layers. It also induced the formation of a substantial amount of sea-ice which nevertheless, did not develop into a permanent pack until the end of the Matuyama epoch of reverse polarity.

Approximately 900,000 years ago temperatures dropped again, low enough this time to freeze the entire surface of the Arctic Basin solid, which then acquired the permanent ''skin'' of ice, 2-4 meters thick, which constitutes the polar pack today.

Needless to say, this is only a simplified view of the rather complex evolution of the Arctic Sea during the Pleistocene. In actual fact, the alternation of foraminifera rich and poor sediments in deep sea cores demonstrates that numerous fluctuations in ice thickness, salinity and nutrient contents did take place during the Mindell, Riss I, Riss II and Riss III glaciations and corresponding warmer interstadials.

At least six times during the last 225,000 years, polar waters expanded southwards and even reached the low latitudes of Portugal, introducing a wide variation into the hydrographic pattern of the North Atlantic and the adjacent Ocean.

Some 50,000 years B.P., world temperature dropped sharply again and the Northern Hemisphere entered into the last glacial period, the Wurm-Wisconsin, an era of bitter cold which became particularly intense 20,000 years B.P.

According to H.H. Lamb, the circumpolar atmospheric vortex developed a great trough southwards over North America at that time, as well as over the Norwegian Sea, Northern and Western Europe. As a consequence, the Gulf Stream was diverted eastwards and the Atlantic flow into the Polar Basin was substantially reduced, if not cut off entirely. Polar waters were again driven southwards to the 30-40 degree latitudes. On the Arctic Ocean a cyclonic circulation prevailed in summertime, resulting in a reversal of ocean currents and ice drifts. This, in turn, provoked upwelling in the central Arctic which led to the paradox that, in the peak of this glacial period, there could well have been an intermittently open sea around the Pole itself.

The main consequences, however, of this new glacial period were the development of massive ice shelves on northern continental lands and, in direct correlation, a huge drop of nearly 200 meters of the mean level of the oceans. Thus the Bering Strait region became a land bridge and all communication by water was cut between the Pacific and the Arctic Oceans. On the Eurasiatic land masses men sheltered in caves and some of these settlements can still be found on the Mediterranean coastline some 150 meters below the present sea level. It was also during this period that ancient man, plodding through the Siberian wastes in the pursuit of big game, made his way east and reached North America via the Bering Bridge.

These conditions lasted for another 9,000 years and it is quite certain that it had a major influence on the geomorphology of the Northern Hemisphere as well as on the structure of its present ecosystems. Finally, some 11,000 years ago, icecaps melted, glaciers receded and a vast oceanic transgression brought the seas back to their present level. However, northern lands and water masses were profoundly marked by this last glaciation and are still slowly recovering even today.

Thus, the Arctic Ocean, identified as a separate physiographical unit only some 2,000,000 years ago and subjected to acute cyclic climatic changes until very recent times, is definitely a young unit. It is therefore not surprising to see that most of its maritime ecosystems are still immature and have not yet reached genetic stability.

A WORLD OF ITS OWN: THE POLAR MEDITERRANEAN

Because of their geographical position, the Arctic regions receive an overall low input from the slant illumination of the sun in summer and radiate vast amounts of energy into the infinite sink of outer space during the winter months. Moreover, on the Polar Sea itself, the permanent pack ice reflects 60-90 percent of the meager incoming radiation due to its high albedo (0.6) and—at the same time—when it is covered with snow, insulates the warm waters from the surrounding air. Under these conditions, the cold air-masses overlying the central part of the Arctic Ocean show a definite trend to subsidence and a permanent High develops over the Polar area.

However, these air-masses are not severed from the global atmospheric circulation. They are the remote northern end of a massive meridional transport, bringing hot equatorial air into higher latitudes and impacting into a vast circumpolar vortex of upper westerly winds. The air masses might develop a fairly smooth zonal distribution or present marked meridional patterns when extended troughs develop in the upper flow.

North of that zone of active circulation, the Polar Highs are quite stable and generate steady easterlies at the outskirts of the permanent anticyclone. However, in summertime, because of the strong thermal gradient which develops between the ice-covered ocean and the bare, dry, snow-free circumpolar continental surfaces, a sequence of deep, mobile, cyclonic storms might appear at the edge of the Arctic Ocean (Lamb).

In the surrounding seas, the main currents result from the prolonged action of the semi-permanent atmospheric streamlines and their given patterns influence the heat balance of the Arctic directly, even from a very long distance. For example, the partition of the Equatorial Atlantic Current at the level of the Northeast corner of Brazil has a direct impact on the heat input into the Northern Atlantic and, via the intermediary of the Gulf Stream, on the heat transfer into the Arctic Ocean through the Fram Strait.

This particular Strait, which lies between Northeast Greenland and West Spitsbergen is, in fact, the crucial point of the whole system and the very place where most of the heat and mass transfers take place between the Arctic Ocean and the surrounding seas.

Actually, due to its particular structure as a semi-enclosed sea, the Arctic ocean is fairly isolated from the other major oceanic compartments. A vast expanse of 9.5 million square kilometers, it is almost entirely landlocked and, with the notable exception of the Fram Strait, communicates with southern seas only through narrow maritime passages of limited depths such as the Bering Strait, the Canadian Arctic Archipelago channels, the Robeson channel (between Ellesmere Island and East Greenland) or via shallow epicontinental seas such as the Barents Sea. Thus, the essential part of the water exchange is carried out through this wide, deep trench (50-100 kilometers wide with up to 2,500 meters depth) where, on an average, 7,000,000 m^3/sec of warm saline Atlantic waters are fed into the Polar Sea, while an equivalent amount of cold ice-covered Arctic waters is discharged southwards along the East Greenland current. This formidable thermal machine is, in fact, a very sensitive detector of minor climatic changes and the measurement of prevailing temperatures and the extent of ice-cover in the areas is a reliable index of the current state of world climate. (The same also applies to the sensitive zone where the northeastern branch of the Gulf Steam becomes extinct: East Barents Sea, Novaya Zemlya and Kara Sea.) It is also an early warning system for ice conditions in the Davis Strait and, as such, its monitoring presents a remarkable interest for the routing of ships bound to Northwest Greenland, the Hudson Bay or the Northwest Passage. This, indeed, is the reason why the Comité Arctique International quite recently prepared a research proposal, called the ''Fram Strait Project,'' with the object of organizing international cooperation for the improvement of our knowledge on the hydrography and dynamics of this most important passage and for a thorough evaluation of its heat budget.

Once the Atlantic waters have entered the Arctic Ocean, they settle at mid-depth and thus begins a long period, over several years, of circulation into the internal sea. The major portion will remain in the Central Ocean where deep basins contain the essential part of the water mass. A small proportion, however, might be edging upon the surrounding epicontinental seas (Barents, Kara, Laptev, East Siberian, Chukchi Seas, and the southern part of the Beaufort Sea), which cover nearly 36 percent of the Arctic Ocean but contain only 2 percent of the total water mass, and where it will remain below the freshwater runoff coming from the mighty Siberian rivers and the Mackenzie. This powerful maelstrom is governed essentially by two main currents: the Transpolar Drift Stream along Eurasia and the Pacific (or Beaufort Sea) Gyre, north of North America. Together with some minor eddies they maintain the whole mass in movement and are responsible for the continuous circumpolar drift of the Polar pack. Thus, in the Arctic Ocean, the surface ice is in permanent motion and this, jointly with the combined efforts of tides and winds, create a colliding and rafting of the ice floes resulting in the development of an often chaotic, tortured surface with huge pressure ridges and corresponding deep ice keels: a sinister and defeating discovery for many early Arctic explorers and, still, a permanent threat for icebreakers and nuclear submarines.

The obvious result of this typical circulation pattern is to give to the Arctic Ocean some kind of an overall solidarity, the effect of which links all individual compartments. As such, the Arctic Ocean is really a ''Polar Mediterranean,'' an internal sea where the specificity of individual sites is somewhat erased by the global behavior. It can be easily understood therefore, that this could have major consequences in the field of environmental control, since the fate of any pollutant released in some localized spots is inevitably bound to be systemic within a certain time. It could also explain why deep concern developed recently over Russian plans to divert a substantial part of the flow of the Ob' and Yenisei Rivers towards central Siberia, an operation which could definitely alter the regular input of fresh water into the Arctic basin and have unforeseeable consequences on its ice cover and heat balance.

This, in fact, is only one of the many aspects of interdependence and cross-linking within the Arctic regions as is also amply demonstrated by the history of human penetration into the northern territories.

40,000 YEARS OF ARCTIC DWELLERS

Ancient man was already expanding into the northern lands as early as the Neo-Pleistocene and we have archeological evidence that, 40,000 years ago, Neanderthal tribes migrated into the cold Arctic tundras. Each single glaciation, of course, pushed them out towards lower latitudes and often compelled them to shelter in caves. However, during the following interstadial they expanded north again and, finally, at the peak of the cold, dry Wurm-Wisconsin period, reached North America along the desolate steppes of the transient Bering Bridge.

Stemming from a common nucleus of Arcto-mongoloids, these primitive hunters further diversified in the New World and, some 10,000 years later, were severed from their Siberian roots by the re-opening of the Bering Strait. Athabascan and Algonkian Indians developed inland while the Paleo-eskimo, ancestors of today's Inuit, spread along the Arctic coastlines or followed the ''musk-ox way'' inland or the main caribou herds' migration routes.

In the vast, deep taiga of Siberia, a similar evolution took place and many individual groups moved northwards into the coastal Arctic tundra and settled from the Kara Sea to the Chukchi Peninsula. Here again, over millenia, they developed highly specialized cultures and were, in the course of history, further joined by other daring adventurers who fought their way up to the Polar Sea. To the west came the Samis and also the more specialized Ugrian populations who gave root to the Samoyeds, the Ostiaks, the Voguls and the Zyrians. To the east, a powerful offspring of the Turks gave birth to the People of the Arctic Horse: the Iakuts, a fierce and mighty wedge right in the middle of the Tungunz population. Still further east, Iukaghiri, Koriak and Kamtchadales spread west and south of the small isolate of Chukchi and Eskimo of Northeast Asia.

By living on sea mammals or by trapping in the sub-Arctic taiga, nomadic or permanent settlers and the Arctic natives conquered the frozen polar wastes thousands of years ago. Remarkably adapted to one of the harshest climates on earth, they managed to survive, expand and tame this desolate universe of ice and rocks. They owned nothing but the land and the sea; their basic resources were theirs up to the remote horizons that encompassed their mobile world. Hunting, fishing, trapping, they lived ''on the fat of the land,'' but they were in such perfect harmony with the environment that they never disturbed the balance of the delicate boreal ecosystems and left no trace of their continuous, albeit transient, presence.

Clustered in temporary huts or igloos, or sheltered in the warmth of their low-profile stone houses covered with hides, driftwood and turf, they fostered familial links and grouped into cohesive communities. They had no leaders, no chiefs, but took great heed of the advice and experience of the elders. Often, in the depths of the polar night, they gathered to hear the magical drums of the shamans, Angakut or Noaids, chasing away the devils and interceding for them with the remote, powerful deities. In the course of the long Arctic winter, their daily life as well as their understanding of the cosmos became imprinted in stores and legends; a living heritage, never transcribed but orally transferred in its vivid, ever-nascent integrity from generation to generation.

Many thousand years have passed but until recent times, it seemed like yesterday that circumarctic populations lived the life their fathers and their fathers' fathers had always lived. They spoke the same language, the language that came to their ears and minds, impregnated with its historical patrimony of legends and traditions.

Is it not easy to understand, therefore, how sensitive they might be today when their once-private domain opens wide to southern enterprise and their native languages and cultures risk becoming extinct under the powerful pressures of Western "civilization"?

A LAND OF RICHES, A LAND OF WEAKNESSES: THE ARCTIC PARADOX

Considered for millenia as forbidden horizons of mystery and desolation, the Arctic regions remained essentially outside the main currents which, since the end of the Middle Ages, drove "modern" entrepreneurs into the wealthy adventures of industry and trade. Despite the occasional "gold rush" of Frobisher to South Greenland in the years 1576–1578 and several unsuccessful attempts to open an Arctic route to China (Cathay) through hypothetical Northeast and Northwest passages, all sea-borne expeditions bearing to the North became invariably trapped in the drifting pack ice and ended miserably.

Even much later kings, bankers and merchants did not consider it worthwhile to fight distant battles to maintain "quelques arpents de glace" ("a few acres of ice") or to develop the huge wealth of the fur-bearing wild animals of the Canadian forests discovered in the "Nouvelle France" by Jacques Cartier as early as 1534. In fact nobody attempted to challenge the natives on their own lands for the very simple reason that there was a complete lack of interest in doing so.

The situation regarding the circumpolar seas, on the other hand, was dramatically different. Indeed, in 1607 the Hudson expeditions to the recently discovered "Het Nieuwe Land" of Wilhelm Barents (1596)—Spitsbergen—showed that in cold northern waters whales abounded by the hundreds, "like carps in a pond." As a result, for the next three centuries, European whaling fleets competed harshly on those "fishing grounds," most unfortunately reducing the initial stock of cetacea to near extinction. At the same time, and even earlier, Basque and Breton captains sailed every year to the rich banks of Newfoundland to return heavily laden with hundreds of tons of salted and dried cod.

Meanwhile, northern land resources, and especially metallic ores and fossil fuels, were completely ignored since southern mines were so much more easily accessible and world needs were still substantially lower than production capacity.

However, in the wake of the great wave of international enthusiasm for scientific exploration and research which occurred in the second half of the 19th century, some thoughtful individuals began to identify large mineral deposits in the Arctic regions. This is how, at the turn of the century, the American, Longyear, discovered thick seams of high quality coal in the Advent Valley in Spitsbergen. Around that same period thousands of would-be millionaires rushed to Dawson City in the remote Klondike or started to dig, day and night, on the beaches of Nome in quest of gold.

Following the industrial development taking place during the first part of the 20th Century and the progressive improvement of both technical and geographic knowledge, the Western World progressively revealed the huge resources of the Arctic El Dorado. Then, the sudden intrusion of commandos, air squadrons, battle ships and large transport freighters during the last World War, peaking with the Murmansk Convoys, definitely opened the remote—and once unknown—Arctic to international enterprise.

Yet the barren, frozen boreal wastes are still fundamentally hostile and almost sterile. Blanketed by snow and battered by raging winds for more than six months a year, they have only a dwarf vegetation providing scarce resources for the very few animals which, like the reindeer, are physically adapted to withstand the Arctic climate and forage for their food in the deep, fluffy snow of the high Arctic. All around the reservoirs or deposits lie hundreds, thousands of miles of desolate lands, crisped by cold in winter or turned into mosquito-infested marshes in summer. There, in the frozen North, is nothing but a long, distressful solitude for those who volunteer to challenge the last Frontier at the rim of the Polar Sea and cannot demonstrate the genetic adaptation and functional ability of the native peoples: a striking contrast between potential wealth and actual misery, a formidable gap between glittering development schemes and still unperfected technologies.

A DIFFICULT BALANCE: FRONT-LINE TECHNOLOGY AND ECONOMIC CONSTRAINTS

The development of advanced technologies to conquer polar areas, outer space and ocean depths is certainly among the most significant achievement of modern times. In fact, in order to deploy their activities in those hostile environments, northern pioneers have been compelled not only to fight technical and economic battles, but also to overcome numerous psychological, emotional and social constraints.

In that perspective, the exploration and subsequent exploitation of Arctic energy resources—if not the most spectacular issue—remains one of the major challenges in which private sectors and governments have been involved in the last two to three decades.

• In the High Arctic, Logistics are obviously of utmost importance and extremely difficult to implement because of the harsh climate, the overall lack of any pre-existing infrastructures, and the huge distances which lie between field operations and supporting bases. Transportation, communications, supply and maintenance of equipment, commuting of crews, medical and health care, emergency procedures. . . have to be developed in each individual case and these often are stretched to the limit.

• The Technical and Human Challenges are also truly formidable in the Arctic environment. Building a 1,300 kilometer pipeline on sensitive permafrost from the Polar Sea to the Pacific—crossing three mountain ranges, more than 600 rivers and capable of withstanding telluric movements and huge thermal stresses—is only one simple example of the high level of expertise required for Arctic resource development. The same would apply to offshore exploration where drilling ships and rigs have to operate in broken ice under the constant threat of drifting pack ice, icebergs and ice islands. Here again, front-line technology had to be developed from scratch; to design and build appropriate structures capable of resisting the formidable ice pressure or to be quickly dismantled and moved away from the site should the towing of major icebergs prove to be impossible or too hazardous. We could also quote the use of reinforced ice-platforms in the Canadian High Arctic, a very sophisticated technique which, for several months, allows a compass point on the sea ice to be converted into a full line drilling and exploration base.

However, strain and stress do not apply only to equipment and structures—but also have impact on the people who have to live an active and difficult life in the most demanding environment. Low temperatures, high winds, extreme air dryness, shifting photoperiods and periodical commuting impose a heavy load on the physical and psychological balance of Arctic workers, and it has been shown that this can induce profound disturbances, not only on sleep patterns, but also on vigilance, mood and neurophysiological equilibrium.

• Under these circumstances, it is not surprising that the Economics of such operations become a critical issue since their cost is at least double in magnitude than at mid-latitudes. It is thus a matter of permanent and deep concern for corporate management or top rank government executives to assess the relevance of such enterprises in a shifting economy, which requires an in-line evaluation of global policies and of their consequences on world trade of industrial commodities.

Arctic development programs in North America, for instance, had definitely a high appeal at the time of the 1973, 1979 oil shocks since they appeared to offer permanent, stable sources of petroleum products in a period of high political instability. Today, overproduction and conservation policies have eased the pressure considerably while, on the other hand, the sensitive Middle East area seems to have settled down into some kind of a dynamically balanced *status quo*. At the same time, most oil-producing countries, having developed substantial downstream interests within the occidental industrial nations, are obviously more reluctant to use economic weapons to enforce their own policy.

However, many indicators show that this might be only a transient situation. Most likely, severe tensions will again develop within the next decade on the world energy market as a consequence of the rising need for petroleum products in the Third World and of the rocketing domestic energy consumption of the oil-producing countries themselves, especially in the Arab world. Thus, within 5-10 years, Arctic energy resources are bound to become highly attractive again. That trend can only be reinforced when we include strategic considerations in the overall analysis, since northern oil and gas are not exposed—in the event of an acute world crisis—to the hazards of long-distance maritime transport.

This points to the evidence that, in defining natural resources, we have to maintain a relativist point of view. Indeed, resource assessment has only a meaning within a given historical context, for a given balance of social, cultural and economic issues and it is strictly dependent upon a given price structure, taking into account that it might be needed to build into this price nonmaterial factors such as national security and "quality of life."

THE BOREAL ENVIRONMENTAL MANDATE: A DELICATE AND EMOTIONAL ISSUE

Any project, whether urban, industrial or resource-related, is bound to interact with the environment since environment introduces multiple hazards and constraints in all human activities that, in turn, might affect its diverse sensitivities. As such, a preliminary environmental impact report is an absolute prerequisite in any development scheme to assess its compatibility with a given site at a given period.

This procedure is not only a compulsory step introduced by the legislative machinery into project management, but it also provides consistency and fairness in the decision-making process. In addition, it gives an adequate early warning of the potential adverse secondary effects that could result from the implementation of the project and thus prevents the considerable additional costs of subsequent restoration.

In other words, in the dynamic interplay between private and public interests, Environmental Assessment is a wise and useful exercise. It is far less costly to resort to appropriate mitigative measures in due time—and secure a harmonious, sustainable integration of the project into existing natural and/or man-made settings—than to try to counteract later, the adverse effects of badly planned development. It has also the notable advantage of involving public opinion in the initial stages, thus preventing the rise of emotional, buoyant, grassroots protests, in the midst of the development phase.

As such, environmental concern should be endemic to the management system and be used as a routine tool at executive level, as hygiene or safety (Larminie). It would then reconcile industry and that loose coalition of old conservation groups concerned with preserving natural amenities, and pollution-control lobbies eager to protect public health and "quality of life" which, together, support the environmental movement (Lake).

Numerous attempts have been made to characterize the different impact categories and give them an appropriate rating within the framework of a selected development. They obviously include physical, chemical, biological and ecological factors that can be cross-related with the different projected activities on land, water, atmosphere or biota. The resulting interactions might prove to have local, regional or even global significance and their effects might be transient or long-lasting, if not permanent. However, far more subtle are the areas of human concerns (economic, occupational, social, psychological or cultural categories) since their relative weights are very difficult to establish. Moreover, they are always highly emotional, and can readily transform the environmental concern, whatever its intrinsic value, into a mere political issue.

This is particularly true for the Arctic regions and there are perhaps no regions on earth where it is more difficult to exercise an objective environmental mandate, based on hard facts and devoid of irrational inputs.

There are several reasons which explain this peculiar situation.

• First, the Arctic regions are, indeed, extremely vulnerable because of their climatic constraints and their low rate of biogenesis.

• Second, any chemical or physical pollution is likely to have longlasting effects due to the slow turnover of both the biomass and supporting media. We can quote, for example, the detrimental effects of fallout radionuclides, stockpiled for more than 10 years in lichen—working as permanent bio-accumulators—and then transferred to man via caribou meat in the Nunamiut population of Anaktuvuk Pass, Alaska.

• Third, the combined effect of climatic stresses, changing photoperiods and wide seasonal variations in energy transfers interact strongly with potential toxicants, often at the sublethal level, and may result in synergistic effects with higher biological aggressiveness. For instance, traces of aluminum or vanadium in natural streams can increase the negative effects of acid run-offs tenfold and can provoke rapid elimination of salmonids.

It is quite obvious that for Arctic ecosystems, insignificant toxic levels or sub-critical biological effects do not exist. This is why such a deep concern developed recently in Alaskan Inupiat, when offshore oil exploration was due to start in broken ice in the Barrow area. Everybody knows how difficult it has proved to control major blowouts at sea: The Ekofisk and Ixtoc examples are present in everyone's mind. Thus, we can easily understand that in view of the present available technical knowledge, coastal populations might be reluctant to see a major oil slick developing in the adjacent waters; however, it does not mean that the whole project has to be cancelled. Industry has already shown on a great number of occasions that it was able to meet new challenges vigorously and to develop rapidly appropriate solutions. There is no reason to suspect its capability to do so and it would be most unfair not to believe in its deep sense of responsibility. However, development takes time and, in the present circumstances, it would be unwise to rush the operations if we wish to ensure both security and efficiency. More than ever, it might be urgent to wait, in this particular case, until further knowledge and improved technology are acquired.

• Fourth, due to its particular geographic position, the Arctic is also, most unfortunately, a pollution sink for airborne pollutants driven towards the high latitudes by the meridional streamlines of global circulation.

Such is the case of organo-chlorinated compounds sprayed in vast amounts in the tropical and subtropical belts and which, by successive distillation-redeposition mechanisms, end up in the fat of circumpolar animals. This is also the case of tropospheric and stratospheric radionuclides, as we have seen above, and for the acid fumes released by coal or fuel-fired power stations or major smelting plants (Sudbury—Norils'k). However, perhaps the most striking phenomenon is the development in wintertime of a pervasive, dense aerosol that forms a thick layer at mid-altitude over Northern Spitsbergen, Greenland and especially Alaska: the Arctic haze. Atmospheric chemistry as well as detailed studies of the air-flow patterns over the North Polar area have demonstrated that this Arctic haze is, in fact, a pollution; an aged air mass of continental origin which finds its source in the industrial areas of Western and Eastern Europe, Western U.S.S.R. and North Siberia. It follows an extensive Eurasiatic pathway, powered by the Icelandic Lows and the Siberian Highs, before being injected into the trans-Arctic circulation near Novaya Zemlya. Its impact on the environment is still poorly understood, though it demonstrates evident radioactive, nucleation and depositional effects which could, in the long run, impede the development of Arctic ecosystems; an obvious case of transborder pollution, it also raises the delicate problem of international liabilities in long-distance transnational transport of atmospheric pollutants.

• Fifth, the Arctic, as we have seen, is the homeland of very ancient populations which have developed, in the course of their long history, specific cultural patterns that are always closely related to the environment. As such, any change in the environment, even if it is felt insignificant at the "expert level," might indeed have deep repercussions on the traditional life of native settlers and could well trigger long and difficult environmental battles.

This is precisely the case of the transportation of Liquified Natural Gas (LNG) from the Canadian Arctic Archipelago to the North American East Coast. In this particular project—the APP (Arctic Pilot Project)—it was proposed to route mammoth icebreaker tankers through the Northwest Passage and along the coast of Greenland to avoid the iceberg drift in the Labrador current. Under these conditions it was forecast that the ships would open broad leads into the Greenland pack. It was also assumed that they would generate in their path a considerable noise, arising mainly from the propellers. The low-frequency sounds produced as a result would be transmitted through the seawater and interfere with whale sonic signals at distances of up to 100 kilometers.

Starting from these unquestionable points, the whole affair was whipped up into an emotional and political issue. Boosted on the international scene, at the European Parliament and the United Nations, it generated such a tense situation that, despite many attempts for intermediate hearings and negotiations, the project was finally dropped in the wake of waning economic opportunties. Whatever the respective positions of the different parties involved, the whole process was somewhat disconcerting. Indeed, much more research would have been required in order to assess with any certainty whether any such ship passage, at rather distant time intervals, would have had a lasting effect on the cetacea and whether the artificial opening of leads into the pack ice, at some distance from the coastline, could seriously impair near-shore hunting. It would also have been useful to put this particular project into perspective with all the other acoustic inputs into the Davis Strait, resulting from regular shipping and military activities as well as from the rafting and colliding of the ice floes.

Unfortunately, a similar situation exists offshore Alaska in the Beaufort Sea where Inupiat populations are afraid they will see oil developments interfere with the migration patterns of the bowhead whale. This animal, which plays a major role in their culture, is generally harpooned from April to June, when it cruises in the open shore lead and is within easy reach of light skin boats powered with outboard engines. The general opinion prevailing among the Inuit is that during this sensitive period, any foreign noise, such as the operation of a drilling rig or the explosion of air guns for seismic profiling, is bound to scare off the whales. This is why they formally requested to forbid drilling and sounding during the migration period, which after all, appears to be a fairly reasonable proposal.

Here again, however, it would be necessary to know more about the behavior of the bowhead whales, their exact route, their sensitivity to sound, their capability of adapting to a changing environment. . . all questions of highly scientific interest but which imply extensive long-term research. Pending results, operating restrictions would, of course, still be applied. Under these conditions, we may wonder whether such restrictions would not finally make the whole project unattractive for major multinational corporations and drive them to transfer their activities to southern latitudes. Would this not eventually, in fact, be more detrimental to the development of Arctic communities than the potential impact on their traditional mode of life? Would not this whole issue represent a serious setback for the State of Alaska and for the Nation?

There is no easy answer when economics clash with tradition and mega-projects have to be weighed against such intangible and sensitive values as cultural identity. Are not industrial development and preservation of traditional values of equal importance?

This thorny problem illustrates more than ever that environmental issues are of a complex nature and that, in most cases, moderation is imperative. The only way therefore, is to proceed step by step and to develop a climate of confidence and good will between the different parties that will foster partnership and not degenerate into litigation. It is a difficult but highly rewarding experience which will require an appropriate setting, free from political and economic intervention, such as the multidisciplinary, neutral, reliable platform of an academic institution.

THE DEVELOPMENT OF AN ARCTIC BROTHERHOOD: NATIONAL AND INTERNATIONAL IMPACT

In the dawn of their nomadic existence, circumpolar people moved freely in the Arctic wastes and settled wherever they felt the geographic environment and proximity of game could support their families. This is how they came to select appropriate sites on the northern rim of the Alaskan North Slope along the migration route of the bowhead whale, or to establish camps on Ellesmere Island close to permanent polynyas which provided open water for hunting seals. In many places, their movements were intimately related to the ethology of their subsistence game. For example, the Samis used to commute from the mountain ranges to the seashore together with the main reindeer herds.

Despite the fact that they developed almost independently, the different native communities presented many common features: linguistic affinities, shamanism, hunting and sledging techniques, and even shared string games, dances. . . . Progressively separated from one another by the inevitable course of history, they nevertheless retained a nostalgia for their ancient past, when the Arctic was a single entity, open to all northern peoples, and this emotion is still present in their minds today. In this respect, let us not forget the major impact of the long trans-Arctic travels of Knud Rasmussen and the deep interest which arose from the publication of the books of Peter Freuchen or the legends and drawings from Aron of Kangek.

Unfortunately, during the last century, an invasive bureaucracy progressively fenced the Arctic people within the national borders of the sovereign states to which they happened to be attached. Moreover, following the October Socialist Revolution, more than half the Arctic became enclosed behind an almost impenetrable iron curtain. More recently, the turmoil of international politics invaded their own lands and unfortunately, sometimes drove them into the maelstrom of war. Charred villages and skeletons of machine-gunned reindeer herds still mark for the Lapps, the retreat of Nazi Commandos from Finnmark.

However, this sometimes tragic evolution brought the Arctic people progressively into the spotlight at a period when, in the great post-war movement for human rights, international organizations commenced their deep involvement in the delicate question of national minorities. Indeed this has been and still is, a most sensitive issue; international law might interfere with national sovereignty in an area that also concerns common ethics and it therefore becomes more a moral problem than purely a legislative matter. We can easily understand the feeling of frustration that could well develop in the Fourth World when they see that many different ethnias of the Third World have obtained access to full political responsibility through independence.

However, this is not a one-way process; we can also easily understand the equally strong bitterness of the majority of the citizens of a given state when they witness the centrifugal aspirations of national minority groups that they have helped, supported and protected for so long from avaricious outside influences without any particular reciprocity.

Obviously, there is no simple solution to this problem and it is quite clear that responsible statesmen cannot sacrifice national unity, nor can they neglect the basic rights of a people to assume its own destiny. It can only evolve by consensus and through patient, elaborate, careful negotiations carried out in a climate of friendship and trust derived from a long, common historical past.

Whatever the outcome, this is definitely bound to be a most important issue in the Arctic and recently, native peoples have made significant moves in that direction. Indeed, we can say that the official "birth" of the *Inuit Circumpolar Conference* (ICC) in 1980 in Nuuk (Godthaab, Greenland), has substantially changed the political climate in the Northwestern Arctic. Proclaiming loud and clear that the "Arctic is our common responsibility," the ICC has rapidly developed into a transnational "Brotherhood of the North" that is quite determined to intervene in any project taking place in the Arctic and even to challenge its administrative structure. Solidly anchored on the Alaskan experience of the Native Land Settlements Act, it supports a similar evolution in Northern Canada and draws its strength and entrepreneurship from the success of the Home-Rule Administration in Greenland under the wise guidance of Premier Jonathan Motzfeldt.

However, the objectives of ICC are not limited to North America and the Northwest Atlantic. It fosters an ambition to include in its field of competence the Soviet Inuit of Northeastern Siberia and also to link its efforts with the Internordic Sami Council. Needless to say, on the first issue, the ICC has met with no success whatsoever since Soviet authorities do not appear to welcome transnational bonds between their population and the Western World. As far as collaboration with the Samis is concerned, it is still rather loose but is building up progressively despite numerous obvious communication problems and substantial basic differences in the structure of both ethnias.

How far it will go is still a matter of personal conjecture and will depend largely upon the attitude and inclination of the present ICC leadership. However, it is quite certain that it will be more and more difficult, if not virtually impossible, to project any development in the Arctic without appropriate interfacing, not only with local populations, but also with the ICC itself, even if the projects concern only a restricted area. Having obtained consultative status at the United States in early 1983, and having been invited officially to Washington to testify during the Senate and Congressional hearings on the "Arctic Policy Bill," the ICC and its executive are firm in their decision to be present as a compulsory international partner in any eventuality within the Arctic.

Quite obviously, the different governments, having competence over Arctic territories, are certainly not enthusiastic to see their authority challenged by transnational pressure-groups even though they respect the moral role and enterprise of the ICC. They have often reacted promptly and taken appropriate steps to release tensions or to solve hot questions before the ICC became involved.

The nomination on December 23, 1983, for instance, by former Premier Trudeau, of the Inuit Charlie Watt to the Canadian Senate (a life-time appointment) has been a clear acknowledgement of his leadership and some kind of compensation for his failure to capture the presidency of the ICC in July 1983 in Iqaluit (Frobisher Bay). Along the same lines, the decision by the Department of Indian Affairs and Northern Development to reject the Stokes Point-King Point development plans proposed by Gulf Canada Resources and Peter Kiewit Sons should be understood as an attempt to link industrial projects to political issues. In fact, this has been stated clearly by Minister John Munro when he declared ''Native groups feel that their outstanding land claims must be settled before large-scale development activities can proceed. While they have never suggested that development is incompatible with environmental management or their own interests, they feel strongly that the issues of claims must take precedence. . . . We all agree that the settlement of land claims will provide a clear, stable framework for advancing social and economic development in the North.''

In the same way, the North Slope Borough has been dealing continuously with the States of Alaska and Washington as well as the private sector for the development of a common policy on oil resources exploration and exploitation without committing the ICC directly or even indirectly. The same would apply equally to the relationship between the Danish authorities in Copenhagen and the Greenlandic Homerule Government.

The ICC, however, is not prepared just to stand by. On the contrary, it appears very much decided to intervene in local affairs. As a case in point, it is interesting to note that on July 24, 1983, the ICC appointed Judge Thomas Berger, a former Supreme Court Justice in British Columbia, to chair the Alaska Native Review Commission (ANRC), which plans to review the impact on Alaskan natives of the Alaska Native Claims Settlement Act (ANCSA) of 1971. While nobody can deny the competence and integrity of Judge Berger, the fact remains that it is most unusual for a transnational native organization to select a Canadian judge to implement an impact study of an American law on Alaskan territory. This could have far-reaching political consequences.

But who can deny that ''since time immemorial, the Inuit have based their lives in the Arctic, sub-Arctic and coastal regions,'' and that these ''areas provided the basis for their existence and culture in close balance between man and nature.'' As such, there might be some pertinence in their strong determination to treat this problem on a global basis, irrespective of historical boundaries.

THE ARCTIC IN THE LIGHT OF GLOBAL SECURITY

Situated on the outskirts of the main populated areas, far from industrial or political centers, the Arctic regions have, until recent times, played little role in global strategy. However, this does not mean that, in the course of history, they have not witnessed any military activity, but this was exceptional and had no particular world significance. For example, when severe tension developed between the United Provinces and France at the end of 17th century, three French men-of-war sailed all the way to Spitsbergen to intercept a Dutch whaling fleet where they fought a naval battle 80 degrees Lat. North, sinking 15 ships and capturing 11 others in Beerenbay (Sorgfjord), Spitsbergen. Then, until the mid-20th century, the Arctic seas and lands were not involved in any major continental or maritime southern conflicts. During the last World War the situation changed, and progressively, northern territories became involved in global strategy for two main reasons.

• First, it was recognized quite early that, in order to carry out military operations with maximum efficiency in the European and Atlantic Theaters (especially airborne missions), it was essential to have a reasonably good meteorological forecast. Since some of the cyclonic depressions and their associated frontal systems deploy from West to East, it appeared of the utmost importance to have meteorological stations as far west as possible; this is why the Germans sent isolated commando groups to Greenland and attempted to gain control of Jan Mayen Island (Island X) where they eventually clashed with Danish and Norwegian patrols.

• Second, at the peak of the German invasion of Russian territory, the U.S.S.R. had an imperative need of massive military supplies for its operations on the Western Front, and was highly dependent upon the so-called ''Russian Convoys'' (also called Murmansk Convoys), shuttling between Iceland or Scotland and the White Sea. In order to avoid the German bombers and raiders stationed in northern Norway, the Convoys took a very northern route, sailing sometimes at the limit of the drifting polar pack between Bear Island and South Spitsbergen where they often met with very serious attacks from U-boats and suffered considerable losses. This explains why Spitsbergen itself, became a position of high strategic importance and why several times, major German battleships, including the Tirpitz and Scharnhorst, sailed the Arctic waters of the Archipelago.

Since that period, the Russians have developed an ''exquisite'' sensitivity regarding the freedom of navigation in the Barents Sea. Indeed, bottled up in the Far East with Vladivostok in the depths of the Japanese Sea, in the Baltic by the Danish Straits and the Sund, and in the Black Sea by the Turkish Straits, the U.S.S.R. could not accept its only ''window'' on the High Seas, the Kola Peninsula, being throttled westward by foreign navies. This issue arose many times in the West-East negotiations and, as early as the end of the War, Mr. Molotov, Russian Minister for Foreign Affairs, attempted—without success—to secure in a somewhat stormy meeting with the Norwegian Foreign Minister, Mr. Trygve Lie, a stronghold on Bear Island.

Today, of course, the development of Intercontinental Ballistic Missiles (ICBMs) equipped with multiple independently targetable nuclear warheads, of cruise missiles of surgical precision and of long-range nuclear-propelled strategic submarines equipped with ballistic missiles, has brought about a substantial change in world strategy and considerably reinforced the military significance of the Arctic regions and especially of the Polar Sea.

Indeed, across the Arctic Ocean, the two major alliance blocks stand face to face, on almost 180 degrees of longitude (more than 3,700 kilometers) and within a relatively short distance for modern weaponry. In the depths of the Arctic Ocean itself, nuclear submarines find the most appropriate theater for their deployment since the deep basins and the permanent ice-cover secure both easy operations and an almost total invulnerability because of lack of appropriate means of detection through the permanent drifting pack. Moreover, and since at the peak of the winter season, the Arctic Ocean always presents at least 5 percent of open water, as polynyas, it provides the submarines with a substantial number of ''firing windows'' through which sea-launched ICBMs can easily reach the whole Northern Hemisphere.

Under these conditions, it is not difficult to understand that both the Warsaw Pact and NATO have considered the Arctic as a crucial strategic area and have built, on both sides, a huge system of early warning and defense, together with important air and naval bases. The DEW line stations, as well as the important installations of Thule (Greenland), Keflavik (Iceland), Jan Mayen (Norway) and of the Norwegian Finnmark correspond contrapunctually to the fantastic demonstration of military power carried out by the Russians in the White Sea and the Kola Peninsula.

It is more than evident then, that the Arctic air-space as well as the Arctic seas are of critical importance and, accordingly, have been heavily instrumented.

If the Chiefs of Staff congratulate themselves on this situation, the increasing military hold on the Arctic does not meet the approval of native populations. On numerous occasions, the President of the ICC, Hans-Pavia Rosing from Greenland, has stressed publicly in letters to the President of the United States and the Prime Minister of Canada, his strong opposition to the massive deployment of military equipment and forces in the Arctic which, according to ICC, should remain nuclear-free and demilitarized. It is true that these repeated protests have had little echo in Washington and Ottawa since the security of both nations is at stake and neither administration is ready to compromise on this vital issue. We can almost certainly guarantee that, in the Soviet Union, similar native initiatives would have met with identical results.

Thus, as long as no other means have developed to provide release for international tensions by open diplomatic negotiations, followed by appropriate reliable controls, the Arctic is bound to become the site of a formidable progressive military escalation in order to maintain a reasonable strategic balance on both sides of the Pole.

Once totally ignored by the remote ends of the earth, the Arctic regions, today, are of growing importance, and present a global challenge to mankind. We can only hope that appropriate means will be found to ensure their development in harmony with their sensitive environment, to the advantage of their native settlers and for the prosperity of all nations.

Arctic Policy Concerns for the Year 2000: Global & Soviet Policy Concerns

Terence Armstrong
Scott Polar Research Institute, University of Cambridge

I would like to thank Dr. Spilhaus and Mr. Maton very much for organizing this meeting. It's the sort of meeting that I like. The sort that I don't like are the ones where you have to submit a full text in advance, and every member of the audience has got that text in his hand, and then your jokes are ruined, your surprises are nullified, and the whole situation becomes much more difficult. In fact, when I face that situation, I feel that I must give a different talk, so that it becomes a strategy of double work, which I deplore.

Well, this occasion is very different from that, and I propose to be informal. There is a little extra advantage that you don't have to write anything out, and I haven't written anything out. So if I appear to digress, just let me.

First of all, I must make clear that I am going to talk about the civil situation and not the military situation. It's already been made abundantly clear that the military situation is extremely important, and I endorse this, but I don't know about it. And if I did know about it, I couldn't tell you about it. So I propose to leave all this out. Others much more qualified will fill in later, I am sure.

I will start by considering the present situation, a kind of baseline idea of what now goes on in the way of marine transportation in the Arctic. Look first at the map of the northern sea route, the Northeast Passage (Map 1), which needs little explanation. The black line is where most of the ships go. Sometimes they go around the north side of some of the island groups, which is not shown there, but most often they don't. One might add to that map a line or two in the top lefthand corner showing that there is also Arctic Ocean traffic between Norway and Svalbard.

On the other side of the North Pole, we have, very roughly, this situation (Map 2). The black lines indicate, of course, the Northwest Passage. The Northwest Passage is a rather complex system of waterways of which there are, in theory, a large number of variants but in practice only one or two.

I would draw your attention also to the line north of Alaska and a small branch line leading off the Prudhoe Bay.

There is some traffic there. On the righthand side of the map, note the line up the West Coast of Greenland to a number of settlements there. The bottom line, through Hudson Bay to the Port of Churchill, is the Churchill Bay Route which has conveyed grain from the prairie provinces to Western Europe these last 50 years.

Now, what operates in this part of the world? A number of icebreakers. Here is the list of Soviet icebreakers of 10,000 shaft horsepower, the ones, in other words, that would normally be expected to be found working in the Arctic (Table 1). There are 20 of them functioning at present. Notice the three nuclears. Notice the four shallow draft icebreakers, and at the bottom you will note that three more nuclears are coming along. This is a very considerable and powerful fleet. Notice perhaps also that they are almost all built in Finland. Only nuclears are not.

Map 1. Main traffic routes through the Northern Sea Route.

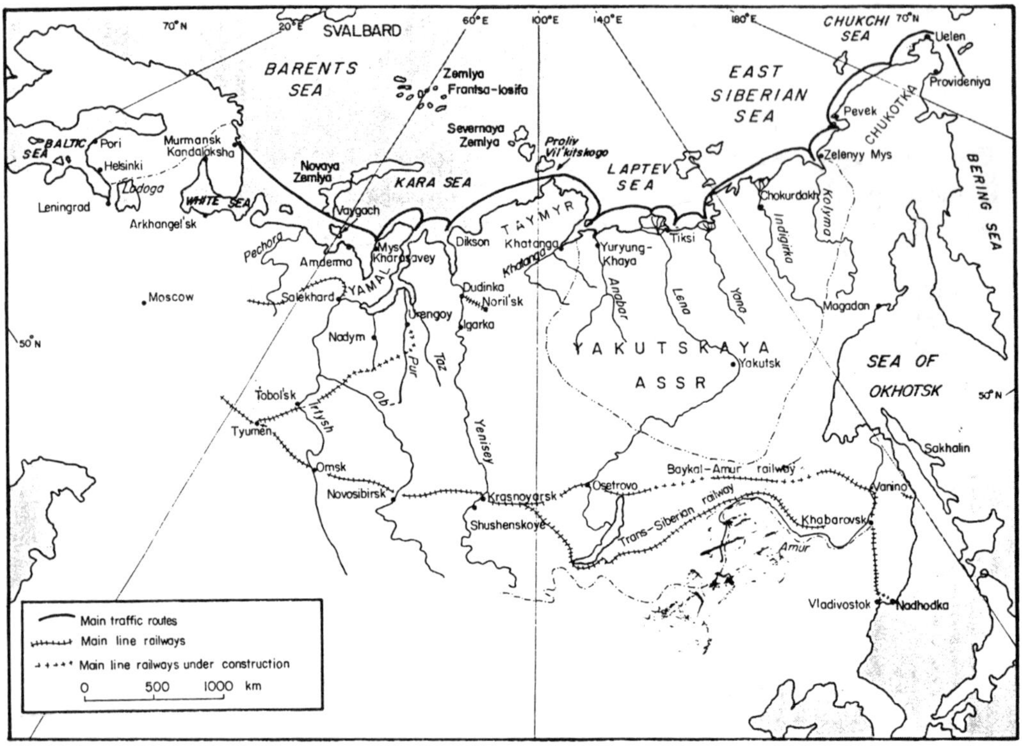

Source. *Polar Record.* Vol. 21, No. 134, p. 499.

Map 2. Shipping lanes in and near the Northwest Passage.

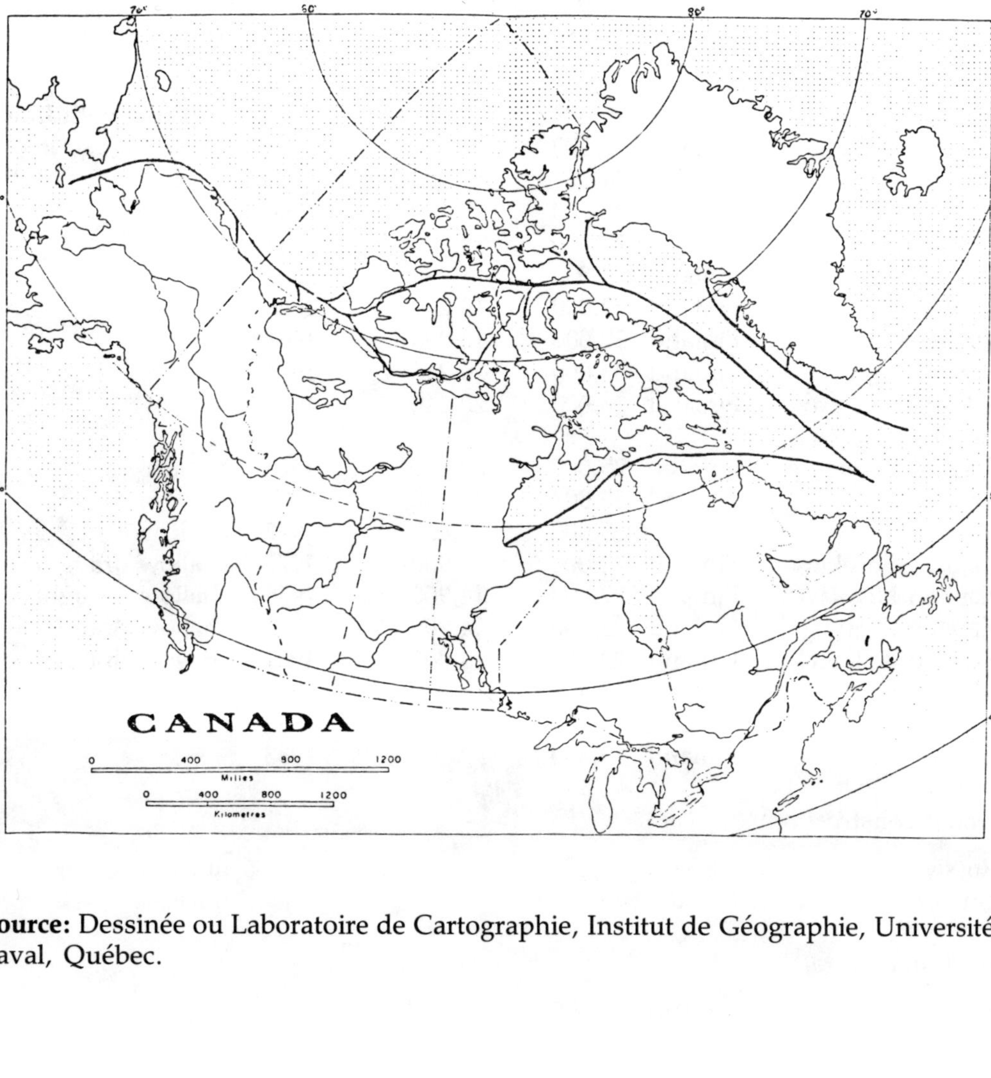

Source: Dessinée ou Laboratoire de Cartographie, Institut de Géographie, Université Laval, Québec.

Table 1. Soviet icebreakers of 10,000 shp, 1984.

Name	Where built	Power shp	Displacement tons	Entered service	Remarks
Kapitan Belousov	Finland	10,500	5,360	1954	
Kapitan Voronin	Finland	10,500	5,360	1955	probably withdrawn
Kapitan Melekhov	Finland	10,500	5,360	1956	
Lenin	USSR	44,000	16,000	1959	nuclear
Moskva	Finland	22,000	13,290	1960	
Leningrad	Finland	22,000	13,290	1961	
Kiyev	Finland	22,000	13,290	1965	
Murmansk	Finland	22,000	13,290	1968	
Vladivostok	Finland	22,000	13,290	1969	
Yermak	Finland	36,000	20,240	1974	
Admiral Makarov	Finland	36,000	20,240	1975	
Arktika, later Leonid Brezhnev	USSR	75,000	23,400	1975	nuclear
Krasin	Finland	36,000	20,240	1976	
Sibir	USSR	75,000	23,400	1977	nuclear
Kapitan Sorokin	Finland	22,000	14,900	1977	shallow-draught
Kapitan Nikolayev	Finland	22,000	14,900	1978	shallow-draught
Kapitan Dranitsyn	Finland	22,000	14,900	1980	shallow-draught
Kapitan Khlebnikov	Finland	22,000	14,900	1981	shallow-draught
Magadan	Finland	12,400		1982	
Mud'yuga	Finland	12,400		1982	
Dikson	Finland	12,400		1983	
Under construction					
Rossiya	USSR	75,000		? 1986-6	nuclear
Unnamed	USSR	75,000			nuclear
At design stage					
Taymyr	Finland/ USSR	40,000		? 1987-90	nuclear, shallow-draught

Source: Updated version of table published in Rey, Louis, ed. Arctic energy resources; proceedings of the Comité Arctique International conference on Arctic energy resources...Oslo...September 22–24, 1982. *Cold Regions Science and Technology,* Vol. 7 (Special Issue), 1983, p. 253.

Whereas, over this side of the North Pole, we have a more restricted list. We've got seven Canadian icebreakers, and we have five American icebreakers (Table 2). I am sure there are people in this room who can correct me and perhaps add new ones or possibly even strike out old ones, but I am rather glad to see that the two wind class are still functioning at the venerable age of 40 years.

Table 2. Canadian and U.S. icebreakers of over 10,000 shp available for use in the Arctic, 1984.

Name	Where built	Power shp	Displacement tons	Entered service	Remarks
Labrador	Canada	10,000	7,000	1954	
John A. Macdonald	Canada	15,000	9,000	1961	
Louis S. St Laurent	Canada	24,000	14,000	1969	
Pierre Radisson	Canada	13,600	7,600	1978	
Sir John Franklin	Canada	13,600	7,600	1979	
Des Groseilliers	Canada	13,600	7,600	1982	
Canmar Kigoriak	Canada	16,400	7,000	1979	
Robert Lemeur	Canada	16,400	7,000	1982	
Northwind	U.S.A.	10,000	5,300	1944	
Westwind	U.S.A.	10,000	5,300	1944	
Glacier	U.S.A.	21,000	8,300	1954	
Polar Star	U.S.A.	18,000 60,000	12,000	1976	gas turbines provide increase in power
Polar Sea	U.S.A.	18,000 60,000	12,000	1978	gas turbines provide increase in power

Source: Compiled from information published in *Polar Record, passim.*

But besides icebreakers, of course, you need ice-strengthened ships, and this table (Table 3) shows you the number of ships on the Soviet merchant shipping register which are classified in the top two of the six categories of ice worthiness that that register identifies. Those two are formally said to be available for Arctic use. There are quite

Table 3. Soviet sea-going self-propelled ships of over 100 GRT in ice categories icebreaker, ULA, and UL, 1980.

	Where registered							
	White and Barents Seas	Far East	Baltic	Black Sea	Caspian Sea	Unattached	Total	No. Built in USSR
Icebreakers	12	10	3	4	—	2	31	18
Specifically strengthened freighters (ULA), chiefly dry cargo	4	12	3	2	—	—	21	19
Timber ships	16	10	6	—	—	—	32	20
Tankers	—	12	2	—	—	—	14	0
Passenger ships	—	6	—	—	—	—	6	6
Dry cargo	8	—	—	—	—	—	8	0
Tugs and salvage vessels	13	37	25	37	17	3	132	63
Container ships	—	3	2	1	—	—	6	6
Hydrographic and expedition ships	9	—	—	—	—	7	16	1
Sealers	9	15	—	—	—	—	24	0
Total	71	105	41	44	17	12	290	123

Source: Abstracted from *Registrovaya kniga morskikh sudov SSSR 1980–81* Leningrad, Izdatel' stvo 'Transport', 1981.

a number, as you can see—290, which includes 31 icebreakers. You may say why 31 when the earlier table which showed only 20. The answer is, there are 11 smaller icebreakers which do not normally, but do sometimes, ply about the waters of the Arctic Ocean.

Now, I haven't got a table to show you the number of ice strengthened ships on the North American side, but I have every reason to believe that it is not as big as that. In fact, I think it is probably very small indeed. I did have the intention, in preparing to talk to you today, to go through Lloyds Register of Shipping and extract them all, but when I found that it was 10,000 pages and there were 40 ships per page, my enthusiasm evaporated, because I think I would actually have found only about 10 ships altogether.

Then there is the question of personnel, and this is where I think the biggest difference between East and West is apparent because, on the northern sea route every year since 1935, there have been at least 50 ships plying, so that a complete career in Arctic seamanship has been possible for the space of two generations. This means that there are many thousands of highly trained and, I believe, very competent Arctic seafarers around, whereas on this side of the pole, it is very much fewer and there is not much chance of anybody spending a long period in an Arctic-going ship.

What freight is lifted? Well, on the Soviet side we don't know. They don't tell us. The most authoritative pronouncement is that by the Minister himself, who was sufficiently concrete as to say "We transport millions of tons in hundreds of ships." He did not vouchsafe how many hundreds or how many millions. I have my own guess, but I don't think I will give it to you now because it may well be wrong.

On the North American side, of course, it is nothing like that. It is, I suppose, of the order of two or three hundred thousand tons, and I stand open to correction.

What is all this traffic? Well, outward bound from the Arctic, it is raw materials principally—minerals and oil; inward bound, it is fuel, food, and so forth necessary to support the settlements where these ores are obtained as well as other settlements. An important point to make is that there is virtually no through-route on either side; even on the northeast passage practically no ships go in from the Atlantic and come out to the Pacific or vice versa. Those that do mostly do it by mistake, because the ice came in and they cannot get out the way they got in.

Well, so much for the present situation in sea transport in the Arctic Ocean and its surrounding seas. What about the future possibilities? What things could be done? Policy options is a rather grand term. I prefer rather to think of things that might happen. Well, one thing, obviously, is that there might be new and better ships. We've already noticed from the table of Soviet icebreakers that three new nuclear ships are planned—two are under construction, and one is at design stage, and there will no doubt be more of those. Indeed, there is the expressed intention of building more of that kind of vessel, not necessarily of that class, but nuclear icebreakers.

Secondly, there is a very important development in the purchase by the Soviet Union from Finland of a class of so-called SA-15 ships, which are freighters strong and powerful enough to be quasi-icebreakers. They have a dead weight tonnage (cargo capacity) of about 20,000 tons and a shaft horsepower of 21,000, so they are in the icebreaker bracket. They are able, or said to be able (and I think it is true), to navigate through a meter of ice unaided and without stopping. This makes them equivalent to a Canadian class III. It is clear from their first year of operation, which was 1983, that they did extremely well, and this fleet will undoubtedly be increased.

There was talk of a 150,000 horsepower ship when the current batch of planned nuclear icebreakers was being talked about, but no action has been taken on that. No doubt people are still thinking about it. The new nuclears, by the way, are thought to be the same as the existing pair, Arktika and Sibir, i.e., 75,000 shaft horsepower.

In Canada, they are arguing about a new icebreaker and have been for some time. Perhaps the Canadians here today will be able to tell me the present state of that argument. There is talk of something a good deal more powerful than anything they've got now, but the actual go-ahead has not been given. Presumably the will to spend that amount of money is not yet present.

In the United States, one may expect replacements for Polar Star and Polar Sea; at least I might expect that, not knowing the American situation firsthand, since those two ships were perhaps less successful than had been hoped. I think probably people will want to see something replacing them.

What about new types of ships? Curiously enough, the basic method of icebreaking is still what it's been for a good many decades: brute force. You simply drive a ship, which is a kind of projectile, at the ice and you hope that the ice breaks first. Generally it does. Other methods of ice breaking have been considered from time to time, for at least 50 years: thermal methods, explosive methods, new forms of bow design, hydraulic methods, but they have not led forward. The only one that has been used, seemingly successfully, is the air bubbling system which is really an ancillary method, a way of trying to inhibit the adherence of ice to the hull of the ship.

What about new types of vessels? A difficulty about using submarines in the Arctic is that immensely broad continental shelf north of northeastern Siberia. This poses real problems. It can be done. I had an American friend in Cambridge who had done it, but I don't think it is likely to be done very much.

Besides submarines, of which we saw some examples in Dr. Rey's slides, there is another possible type of vessel: a semi-submersible. One has been dreamt up by Mr. Pikul of the Soviet Union and made available only earlier this year. The semi-submersible idea is old enough. The hull is under the water but a little control tower is above. What intrigued me was a cockscomb effort on the top of the hull. This is supposed to soften up the ice before it actually hits the structure that supports the control area of the ship. I doubt if that has much more future than any other semi-submersible.

So much for possible ships. Lots of people here know more about these than me, and I hope they will talk about them in later sessions. What about lengthening the season? This is something that is mostly related to ship improvement, obviously. The more powerful the ships, the more ice you can break and the later you can do it (or the earlier). But also don't let's forget sea ice research, which plays a part here, too. If you can actually know how the ice is behaving and avoid it, you do even better than breaking it. But the Russians have had the most success in this question of lengthening the season, because their season in the southwest Kara Sea is now, they say, year round—or almost year round. They can't do the period of ice break-up, but for at least ten or ten and-a-half months, that route is operational. They are doing extremely well, and the question now is whether that experience can be extended to the rest of the northern seaboard. It's on the agenda for a Soviet side, and they hope to be able to do it in the 1990s. I'll say some more, perhaps, a little later about how likely I think they may be to get that.

How close are we to the idea of being able to go anywhere, anytime in the Arctic Ocean? Well, I think actually we are quite a long way. No ship has as yet been through the most difficult parts of the Arctic Ocean, which are in the Pacific gyre area where the ice is moving slower and therefore tends to be thicker. Somebody calculated some years ago, I think in the context of Antarctic ice breaking, that you would need 210,000 shaft horsepower for a ship to go anywhere anytime, and nobody so far has thought of building that.

The next future possibility is what about freight developments? In other words, what's it all going to be used for? Well, you can have, first of all, more of the same—more coastal shipping, more transport of oil, more development of what already happens. This is obviously linked to the development of mining projects onshore, and it is undoubtedly what the USSR will do.

There has been competition, and still is, between the northern sea route in the Soviet Union and the railway to the south. Either you take goods along the sea route and up the rivers or along the railway and down the rivers, and these two have been acting against each other, so to speak. The railways have been winning, but the sea route has not been losing in absolute terms, because although it is getting a smaller proportion of the traffic, its total freight tonnage is probably growing all the time.

But there is now reason to believe that the rail-river route is not able to cope with the demand. This is especially true of the Yana River and the needs of the Yakut Republic. There have been articles in the press in the last year or two showing that the river is now used to capacity. This is related to river depths and the fact that when you want to use the river most is the time when the water is shallowest, especially in the upper reaches. So there gets to be a moment where you just can't use it anymore to the extent you want, and that's where the sea route may well come in.

Then there is the question of major offshore oil development. The Beaufort Sea you all know about, of course. That is now happening there. The Barents Sea is also the site of drilling. The Kara Sea is going to be next on the list. I believe, but other British specialists don't, that they already have started exploratory drilling there. There will also be other seabed exploration and exploitation, no doubt. The question of coping with large amounts of floating ice is going to be raised later by other people in the workshops sessions. It will be interesting for me to hear what is said.

Then, these two routes, the northeast and the northwest passage, could become part of the world shipping lane system. This is something shown on Map 3. It shows the difference in distance by going from Western Europe to, say, Japan, by either the northeast passage or through southern seas; and the interesting point to note is that the breakeven point, if you want to go from, say, Hamburg to the northwest Pacific, is about Hong Kong. If you aim north of Hong Kong, then it is shorter to go by the northern route. This might open up another possibility. There is no sign yet of that happening, but it is something that could happen if the ice problem is mastered to a sufficient extent.

Then finally, there is the question of central Arctic Basin possibilities. The idea was started off by the remarkable voyage of the icebreaker Arktika in August 1977 when

Map 3. Sea routes from Europe to the Pacific, by northern and southern seas.

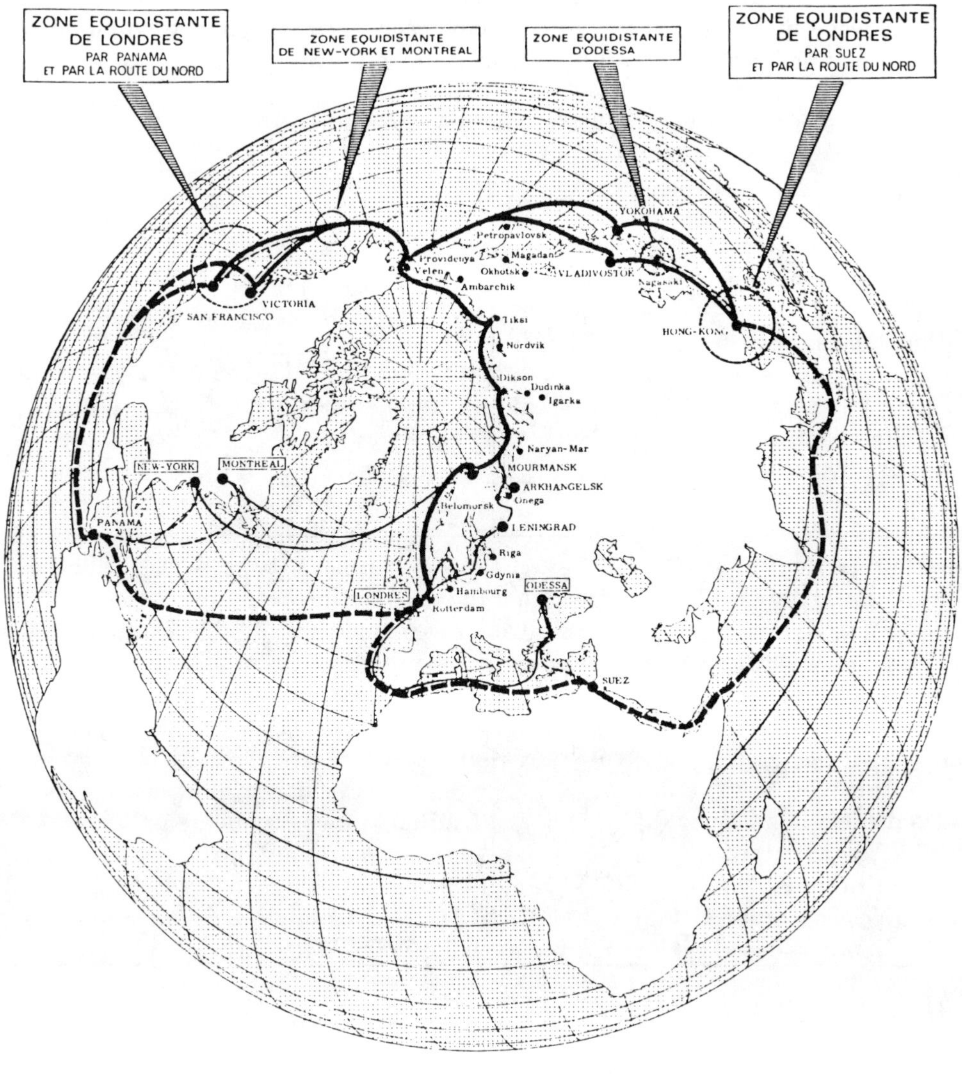

Source: From A.S. Svendsen. La route maritime du nord. Son importance pour le transport maritime et la navigation internationale. Paris, Ecole Pratique des Hautes Etudes, 1963 opp. p. 7 *(Centre d'etudes Arctiques, Contributions, No. 2).*

it went to the Pole and back (see Map 4), a voyage of 3,852 nautical miles done in 14 days at an average speed of 11 and-a-half knots—a staggering achievement. This was the first time a surface vessel went to the Pole.

This voyage was stated at the time to have been done with a view to pioneering a possible trans-central Arctic route to lead to further voyages. And it did.

Map 4. Tracks of Soviet high latitude voyages, 1977 and 1978.

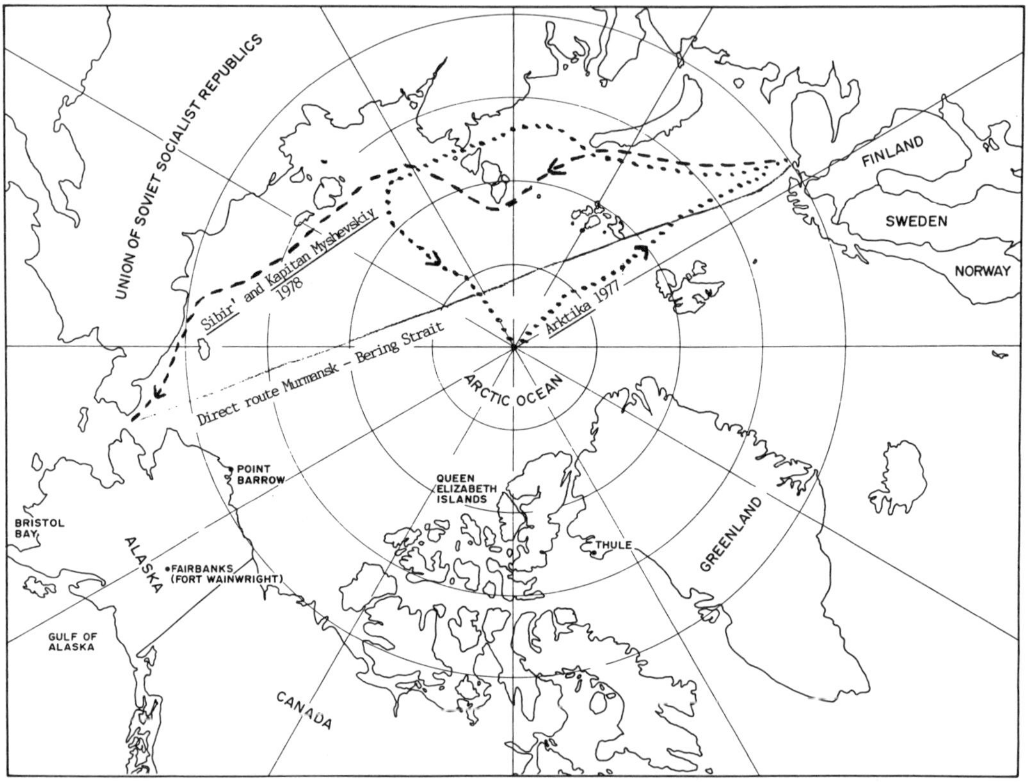

The next year, 1978, two ships, the Sibir, sister ship of Arktika, and a freighter, Kapitan Myshevskiy, followed a course closer inshore, nowhere near the Pole, but north of the island groups, in May-June, much more difficult than August. The voyage was successful, the Myshevskiy got to Bering Strait, but since then, we have heard no more of these voyages.

Now, the point is that there is really no Soviet need to have a route of this kind. The Soviet need is to get into the rivers of northern Siberia and get the ores out. A trans-Arctic route would not be of particular benefit. If the Soviets were using the through-route, then it would be beneficial because it is shorter, as you can see. The direct line is marked on Map 4. It is actually 1,300 kilometers shorter than the coastal route. So if you could offset the ice disadvantage, you'd be gaining. But there is at present no special traffic for which that route would be useful.

I believe that these high latitude voyages were made by the Soviet Union simply in order to demonstrate her own dominant position in Arctic Ocean navigation. For this was the time when Polar Star and Polar Sea were being built. One was afloat, I think, and the other was not. I think that the Soviet Union looked across the North Pole, and people thought, "They are building ships the power of which is about the same as our nuclear icebreakers. If we don't look out, they will beat us to the Pole, and it will no longer be clear that the Soviet Union is the master of Arctic Ocean navigation." Then it became clear that Polar Star and Polar Sea were not indeed likely to get to the Pole, and I think that point was appreciated in the Soviet Union, and priority for further trans-Arctic voyages was therefore lowered, and we haven't heard anything more about them. But if such a route existed, then I think a use would be found for it. Whether it would become a world route or not I don't know, but it is possible. You may know that the Soviet Union offered to throw open the northern sea route to foreign traffic in 1967. Nobody actually took the offer up, and then a few months later, it was, one believes, tacitly withdrawn. This was probably tied up with the fact that there was then a war in the Middle East, and the Suez Canal was closed, and maybe an alternative to the Suez Canal being opened at that time might have been an affront to the Arab friends of the Soviet Union. That is, again, my conjecture. But that is something perhaps for the future.

Who is going to pay for all this? What is the motivation? Well, trading routes in the West, of course, have to be economically viable. So, it all turns on whether the resources being exploited are wanted sufficiently badly. But in the USSR, this need not be so, although I would make the point that it probably is so to a greater extent than you may think. The year round operation of the Kara Sea route into the Yenisey River, to the town of Noril'sk, the nickel mining center, is paid for and made possible by good money produced by the nickel mining combine. This information we have from Mr. Chubakov who is the head of the Northern Sea Route Administration.

So, there is this kind of backing in the Soviet Union. They have to pay their way to some extent, but at the same time, the state pays for massive overheads. I suspect, for instance, that all icebreaker operations are paid for by the state.

Another point that I might make for the future is that expertise might be sold to foreigners. After all, in 1967 they tried to do just this. They would have charged foreign ships a certain fee to be taken through by Soviet icebreakers. That is something that might perhaps come again, and we know also that the Soviet Union has tried to buy expertise in the Arctic from foreigners. There were talks, for instance, with British Petroleum about the possibility of offshore oil work in ice-filled waters.

A word about international cooperation. It has already been mentioned by Dr. Rey. It is highly desirable, in my opinion, that there should be some international cooperation in the Arctic Ocean area. We have the Antarctic Treaty which has worked extremely well for 25 years, although it is under attack now. Something of this kind (actually it would have to be considerably different), something uniting the countries most interested and most concerned with Arctic Ocean navigation would in my mind be a useful thing, and this might start, perhaps, in the area of regulations about pollution control, about environmental control.

But, and this is the great "but," what about Mother Nature? What does she think about all this? At present, the lead of technology over nature is indeed marginal. The ice situation has unquestionably deteriorated. 1972 and 1973 were bad years. 1978 was the worst year ever known. So was 1979—in other words, it was even worse. 1981 was very bad. In 1983, as you know, 51 ships were caught in the ice north of Chukotka.

People who extrapolate climatic predictions from historical data think things are going to get worse. Take, for instance, Willy Dansgaard and his work on the ice core from Camp Century in Greenland. He has been attacked on statistical grounds, but the Russians agree with him. They (N.A. Volkov and V.F. Zakharov) argue from quite different premises (sun spots, ice behavior cycles) and have concluded that things are going to go on getting bad into the 1990s and possibly beyond, and that by the end of this century things will be about as bad as they were at the beginning of this century. Let me remind you that if you look at climate in this part of the world over the last 700 years, the first 600 of those years were worse than this last 100. So it looks as if we are heading for a situation which is going to be worse than anything we've known about while ships of any size have been going into the Arctic, and if it goes on getting bad, then it gets into areas of badness, so to speak, that we know not of.

There is also the question of whether man will play a part in this, will precipitate the change by doing things which have unfortunate results, and the one everybody thinks of is the question of river flow change. What if the Siberian rivers are blocked off? I don't want to get into this hassle, because it is a big argument, and it's great fun, and a lot has been written about it, but I can't start it now. My own personal belief is that it won't matter, but very few would agree with me. If you hold back a lot of water that is going into the Arctic Ocean, you could alter the ice situation, and it could be for the worse. It could also actually be for the better, but that's by the way. So all this means more ice research is needed. The chief Russian concerned with these things, Yevgeniy Tolstikov, has said just that when commenting on the 1983 debacle. He said that what is needed is more ice research, and since the Soviet Union does more ice research than anybody else, then I am interested to hear this statement.

What about the situation in 2000? This is obviously the time when coats should be trailed, necks stuck out, limbs gone out on and so on, but actually, my conclusions are rather tame. Sorry about that. I think there will be a lot more ships built. As I've indicated, I think the Russians will have at least ten more icebreakers. They will have at least 20 or 30 more UAL ships (those are the SA-15 type ships). They may be achieving year-round operation on more of the Northern Sea than just the Kara Sea. Meanwhile, the United States, by the present look of things, may have two or three more icebreakers, and Canada, by the present look of things, may still be arguing. Canadian will put me right.

The trans-Arctic route may be starting. I think that it is something for which the technology is not here yet, but is coming along. It all depends, in other words, on how much money is going to be put into these developments. Submarine freighters will almost certainly be with us, because the technology is all there. It is just a question of building them with the right spaces inside. The ice threat probably will, in my opinion, get worse. This means that there will be more need to spend money on the technology to offset this. This means that there's got to be a greater incentive to spend that money. Who is going to have that incentive? Well, North America, it seems to me, has the healthiest, some might say most voracious, appetite for raw materials. I think that it is likely that the money will be forthcoming from here.

Discussion

DR. ZALESKI: First I think it is a little surprising to read on one of the slides Dr. Armstrong has presented that the Russians are maintaining four icebreakers in the Black Sea. So what do they do with this icebreaker in this quiet, little known—

DR. ASSUR: Adrian Bandersket, he was writing about ice in the Black Sea.

DR. ARMSTRONG: Quite right. Dr. Assur has it. My one sentence answer is, there is ice in the Black Sea.

DR. ZALESKI: I don't think so. Second question is, I appreciated the slide in which you showed submersibles supporting the shocks, but I am not convinced how such floating vessels are stable structurally once the floater is below the water. Is the ice pack considered to contribute to stability?

DR. ARMSTRONG: I don't think so. You'll have to ask Mr. Pikul what his answer is, but this would be, of course, a considerable problem. I think the buoyancy question is very tricky. You've got to have neutral buoyancy.

DR. SONU: One thing that probably escaped comprehensive attention in the Arctic Sea has already happened. It so happened last month we had a team of scientists report on the Beaufort Sea affecting ice flow tracks. On September 13th, we were stranded on one of the offshore islands, fogged in, and suddenly the fog broke up, sunlight beamed down, and a cruise ship appeared out of nowhere, but this ship was cruising from east to west. The next morning, we saw in the newspaper from Anchorage that this cruise ship apparently had left Sweden, gone across the Beaufort Sea, through the Bering Strait and then anchored. It nearly astounded us. One of the most beautiful, pristine areas probably left on earth that is now open to tourists!

DR. ARMSTRONG: Indeed.

DR. SONU: They have a second cruise in May. They almost barely made it this time, because two weeks after that we had a major storm—the worst in 25 years.

DR. ARMSTRONG: I believe for only $20,000 it can be yours, too.

DR. SONU: Just wanted you to know that this is happening.

DR. ARMSTRONG: Yes, very good point, Andrew?

DR. ASSUR: I don't know a better man than Dr. Armstrong to answer the question that I am going to ask, but first a little background. You did mention that the Soviets offered to open the northeast sea route a number of years ago. They announced it will be open for shipping by all nations. They offered icebreaker assistance, obviously for a fee, and then because of political considerations, those people were called back. You probably know Captain Maynagashev and so forth. Now, it just happens that the technology is there or it is virtually within reach of most sites of the Arctic Ocean. On this side, I would mention that we are quite superior in identifying what is going on in the Arctic Ocean by remote means, and satellite observations and so forth. We can see a single ship struggling in the ice. Therefore, by international cooperation, it would be possible to open up not only the Northeast passage but also the Northwest passage for the type of routes you had in mind.

Also, where we are at the present time, we can predict where ice will converge or diverge. The techniques are there, just get it down on a computer.

Now, I am going to ask the question: In your best estimate, do you think this will materialize before the 21st century?

DR. ARMSTRONG: Yes.

MR. MATON: Well, thank you very much, Dr. Armstrong.

ENGINEERING AND ENVIRONMENTAL FACTORS IN THE ALASKAN OFFSHORE

W. M. Sackinger
Geophysical Institute, University of Alaska

The waters adjacent to Alaska have supported commercial fishing since 1864, and this industry has evolved to an annual level of over $400 million employing—directly and indirectly—several thousand people. With other uses of the oceans looming on the Alaskan horizon, such as oil, gas, and mining, there are apprehensions in the minds of present users about the possibility of disruption of their fishing by these new operators offshore. Yet, economic benefit for the nation is seen as a likely result of exploitation of these offshore minerals, particularly in the instance of oil and gas recovery. An overview of economic levels and nations involved in the Alaskan Bering Sea fishery shows that Japan, the Republic of Korea, the Republic of China, and West Germany, as well as the United States are taking several species of groundfish, primarily pollock, Pacific Ocean perch, sable fish, yellowfin sole, Greenland turbot, and Pacific halibut. Most of the Bering Sea falls within the 200-mile Exclusive Economic Zone proclaimed by the United States, but international agreements have opened this fishery to the other nations above, and a 1983 foreign harvest worth $331 million has been reported. Regulations, quotas, monitoring, conservation, and management is fundamentally a federal responsibility for the zone from 3 miles to 200 miles offshore; the North Pacific Fisheries Management Council is responsible for planning in this region. Few Alaskan fishermen seek groundfish, but many salmon (averaging 10 million per year) and shellfish are taken by Alaskans, and it is expected that a major U.S. groundfish fishery will develop in the Bering Sea to gradually displace the foreign harvest. The long-term prognosis for the Alaskan fishery involves a sustained average annual production in the $500 million range, with several thousand U.S. citizens involved.

At the same time, there are several thousand Alaskan natives residing in villages of the Bering Sea coast, harvesting their food for subsistence from traditional species, including salmon, crab, seals, walrus, ducks and geese, sea birds, whales and even polar bears. The use of these species is interwoven into their cultural traditions and village social structure, and is not to be regarded as strictly a matter of economic necessity, although that factor is often present.

Naturally, any change in regional activities (which is perceived as having a possible negative effect on either the fishery or the native subsistence traditions) should be expected to be resisted by residents of the region. In fact, many residents perceive the future development of Alaskan offshore oil and gas resources as a probable disturbing factor for them. The subsistence traditions extend along the coast of the Chukchi and Beaufort Seas as well, although the dollar value of commercial fishing activities diminishes rapidly north of the Bering Strait.

The inevitable problem in the minds of most of those that harvest the living resources of the seas of Alaska is the small but finite probability that oil operations will not always have purely localized effects (such as at a drillsite) but rather that occasionally they may have geographically-extended effects (such as an oil spill) and that such events may harm the living resources of the sea upon which these people depend. Calculations have been made of the probability of such events, the geographical and temporal extent of their effects, and the level of impact upon the living resources of the seas. It is fair to say that the scientific sophistication brought to bear upon this problem has been at the highest level available today, but the assumptions used in any analysis are always open to challenge, and for some Alaskan natives, the only acceptable probability for environmental damage from any human discretionary actions, should be zero.

It is in the light of this regional concern that one must consider the development of Alaskan offshore oil and gas resources from the national and global perspective. The geologic potential for oil and gas in the Alaskan continental shelf regions has been estimated (National Petroleum Council, 1981) at a mean value of 17.6 billion barrels of oil and 74 trillion cubic feet of gas, worth approximately $750 billion at today's prices. If these resources are extracted over a forty-year interval, a revenue stream building up to a maximum value of $30 billion per year could be expected, for petroleum-based products used directly or indirectly by everyone in the entire United States. It is important to recognize the policy dimensions of the Alaskan offshore oil and gas resources, which are: 1) resource use by all Americans; 2) stability of a domestic source of oil and gas supply; 3) a time frame of thirty to forty years for extraction; 4) an annual revenue flow in the $20 billion range; 5) uncertainty of discovery locations, times, and quantities. Traditional use of private capital and the free-enterprise system in the American oil industry has included the custom of oil lease sales, followed by exploratory drilling and production where discoveries warrant. This traditional method of domestic oil and gas production has been used in America for 120 years, involving the activities of hundreds of thousands of people in oil-related businesses. It is the confluence of geological, geographic, socioeconomic and cultural factors since 1865 that have led to today's attempts by two separate industries to operate with minimum conflict in extracting resources from the same region—the Alaskan offshore.

Areas of concern and possible conflict are oil spill effects on fish and marine mammals; acoustic disturbance of marine mammals; modification of coastal ecosystems due to gravel causeways; and economic implications of oil development activities in local cities and villages. On the positive side, the concurrent growth of fishing and oil industries will offer employment opportunities, as well as increased federal, state and local leasing and tax revenues.

Environmental conditions of ice and sea floor provide new challenges for the oil industry in the Alaskan offshore. The solutions to engineering problems involve application of known techniques, in new and different proportions, and with a consequent new set of estimates for operational reliability. Whenever engineering projects are undertaken, the compromise decisions of reliability in relation to cost must be made, and it is such discussions, involving industry designers, government regulators, and regional protectors that are often the most difficult to bring to an acceptable conclusion.

Specific conditions of the environment which affect engineering choices are annual sea ice of nominal thickness (1-2 meters) ridging to heights of perhaps 23 meters and depths of 25 meters near shorelines and fixed structures. Also moving ice exerts substantial lateral force on fixed structures, but ice also is an impairment to marine navigation. Both of these effects of ice can be overcome in the Bering Sa, but the ice of the Chukchi and Beaufort Seas must be regarded as a very serious challenge. Multi-year ice ridges with keel depths to 47 meters, (Lyon, 1967), ice islands with thicknesses of 50 to 60 meters, and large consolidated rubble fields of multi-year ice have been recognized as hazards there. Maximum values of uniaxial compressive strength have been reported as 12.68 MPa (1830 1bf/in^2) for multi-year pressure ridge ice samples (Cox et al., 1984). However, the same study has shown minimum values of uniaxial compressive strength of multi-year ice samples at 0.47 MPa (68 1bf/in^2). If one recognizes multi-year ice as a statistical ensemble of at least eight different categories of ice, (Richter and Cox, 1984), which will fail along planes of weakness despite the presence of strong inclusions, then the calculation of forces exerted upon offshore structures by fracturing multi-year ice becomes a complex problem which should include temperature profile of the ridge, details of the spatial distribution function for the several ice types, and especially the depth of the consolidated layer. Observations of multi-year pack ice in regions of mesoscale pack ice tension show that thick floes can crack in tension under conditions of very low pack ice stress. Evidence for multi-year ice floes 10 meters to 15 meters in thickness which cracked in tension has been accumulated. The stress levels which cause this behavior are obviously low, but have not yet been measured.

One parameter which has often been used to calculate forces on structures in regions of multi-year ice is the thickness distribution of multi-year ridge keels. In the shear zone, however, annual ice is continually refreezing in open leads, giving rise to a condition of annual ice which is actually thinner than in the landfast zone. Forces on structures involved during the fracture of annual ice in the shear zone are low, because the annual ice serves as a fragile buffer material, in most cases, between a bottom-founded structure and the strong multi-year ice inclusions and ice island inclusions within the composite ice pack. Greater forces arise when the annual ice is absent, or has been converted into annual ice ridges, and a significant amount of multi-year ice is available over a large region to apply forces upon a structure. Weather conditions which drive the pack toward the shoreline for several days continuously, for example, could cause this situation. Very old and level multi-year ice from Nansen Sound has been observed to have a thickness of 7 meters (Sackinger et al., 1984) and in another coastal location, 10 meters (Sackinger et al., 1984). Multi-year ridges are often thicker (up to 64 meters) (Reimnitz et al., 1984), but are embedded in multi-year floes of variable thickness.

Of considerable interest is the consolidated thickness of the old multi-year ridge keel. Consolidation involves summer melting of ice in the ridge sails, which drains down into the ridge core until it reaches the surface where negative temperatures prevail and refreezing takes place. The spaces and channels available for draining the meltwater into the ridge core gradually close themselves by this refreezing. Voids in the upper part of the keel are gradually filled with ice, until they are closed and lateral drainage of ridge sail meltwater into surface melt ponds becomes dominant. Of course, when meltwater initially does penetrate into the ridge keel, some lateral spreading takes place, broadening the consolidated region of the keel. With such spreading, it is difficult to see how a very deep keel could become consolidated at its bottom. It should also be appreciated that the core of a large ridge remains near the average annual temperature of the region, which is below 0°C, and the summer impulse of heat applied at the surface cannot penetrate more than a few meters into the ridge core (once convection, of meltwater is negligible), before summer melting is over. Such qualitative arguments lead to the conclusion that there is a limiting depth of consolidation in a very old multi-year ridge, below which there will be a very gradual formation of thin layers of new sea ice (of low strength). Data presented by Kovacs (1983) show that this takes place at a depth of 7 meters and 5.5 meters, while additional data (Voelker et al., 1981) show this at 5 to 12 meters. On the basis of some 28 ridges, it appears that a maximum consolidated thickness of 12 meters is a reasonable preliminary value for purposes of structure design. Further research on multi-year ice ridges and ice properties is clearly very important.

The thickest category of ice hazard in the Alaskan Beaufort Sea is the ice island, at 45-50 meters thickness. These are calved from floating ice shelves along the coast of Ellesmere and Axel Heiberg Islands, and seem to be of rather uniform composition and thickness, although there is little data to support this suggestion. Production of small ice islands takes place continuously each year to some degree (Sackinger et al., 1984), but the larger ones (9 to 35 km in length) are released only infrequently. At least ten small ice islands are presently drifting within the ice pack north of Ellesmere Island. As they progress around the Arctic Ocean, there is a finite probability (which remains to be quantified) that they will interact with a fixed offshore structure.

Because of the existence of the trans-Alaska pipeline, it is unlikely that marine shipment of offshore oil from the north coast of Alaska will take place. The economics of marine transport of natural gas from the same region, however, need to be periodically re-evaluated because there are no obvious insurmountable technical barriers confronting this option. In the Bering Sea, marine transport of oil and perhaps natural gas is the most economically favorable situation (Padron et al., 1984). Of course, environmental constraints of ice forces, wave forces, winds, currents, storm surges, visibility, seismic excitations, and spray-ice accretion need to be taken into account on a site-specific basis in making designs for oil and gas extraction in those areas (Weeks and Sackinger, 1981). A detailed atlas of environmental constraints is beyond the scope of this paper, but it is possible to point out those which are of primary concern for some selected Alaskan offshore activities.

As mentioned above, offshore structures must withstand annual sea ice in the Bering Sea and multi-year sea ice in the Chukchi and Beaufort Seas. Icebreakers, offshore structure deployments, pipeline laying operations and offshore loading operations must withstand ice hazards as described above.

If there are any oil spills, the conventional methods of oil confinement and recovery, using booms and skimmers, should be modified to work in light ice cover. However, in continuously moving, broken ice, new approaches to oil-spill technology must be devised. Oil encapsulation in freezing ice or moving ice should be expected except during spring breakup and summer. The confinement and subsequent treatment of ice which contains oil is difficult, as many approaches are rejected because of danger to human life. Igniters have been used to burn oil after it becomes accessible. It should be recognized, however, that nature has in fact provided a dispersal mechanism for oil spills in the form of ice, which carries oil great distances, releases it gradually into the environment, and finally melts and disappears into the sea. In fact, if dispersion of spilled oil is the objective (as opposed to concentration of spilled oil) then it would be difficult to suggest a dispersant that is more effective and benign than ice itself.

The fishing industry and the oil industry both must contend with spray-ice accretion on boats, ships and structures. Coatings which reduce spray-ice adhesion would be very useful, as they could be applied regardless of surface geometry and would not require energy for de-icing. In the fall, near the ice edge, spray-ice accumulation is very probable, and many design suggestions have been made (Jørgensen, 1982). This hazard must be recognized and taken into account.

In conclusion, one may consider Alaska's offshore area as a vast resource of seafood for the world, as a developing petroleum resource region which may supply perhaps one fourth of the nation's needs, and as a bountiful region of marine life supporting the existence and cultural traditions of thousands of native people. Global driving forces, such as population growth and fossil energy demands, are changing the level of human activity in Alaska's coastal regions from historical subsistence to aggressive exploitation. It is indeed a challenge for the American social system to combine engineering and environmental inputs, and act through the industrial and governmental sectors from international levels through national, state, and local levels, to manage this change in an optimal manner. Arctic engineering in the 21st century must play a major role in meeting this challenge.

REFERENCES

Cox, G.F.N., J.A. Richter, W.F. Weeks, M. Mellor, (1984). "A Summary of the Strength and Modulus of Ice Samples from Multi-Year Pressure Ridges," Proceedings Third International Offshore Mechanics and Arctic Engineering Symposium, ASME, New Orleans, pp. 126–133.

Jørgensen, T.S., (1982). "Influence of Arctic Accretions on Activity in the Northern Part of the Norwegian Continental Shelf," Report STF88 F82016, The Otter Group, Trondheim N-7034, Norway.

Kovacs, A., (1983). "Characteristics of Multi-Year Pressure Ridges," Proceedings of Seventh International Conference on Port and Ocean Engineering Under Arctic Conditions, Helsinki, pp. 173–182.

Lyon, W., (1967). "Under-Surface Profiles of Sea Ice Observed by Submarine," in *Physics of Snow and Ice,* International Conference of Low Temperature Science, Hokkaido University, Sapporo, Japan, Vol. I, pp. 707–711.

National Petroleum Council, (1981). "U.S. Arctic Oil and Gas," Report to U.S. Secretary of Energy, pp. 18–19.

Padron, D.V., E.H.Y. Han, O.P. Richey, M.T. Faeth, W.M. Sackinger, S.G. Sheps, (1984). "Evaluation of Bering Sea Crude Oil Transportation Systems," Report to Minerals Management Service, Han-Padron Associates, New York.

Reimnitz, E., P.W. Barnes, and R.L. Phillips, (1984). "Geological Evidence for 60-Meter Deep Pressure Ridge Keels in the Arctic Ocean," Proceedings of International Association of Hydraulic Research, Ice Symposium, Hamburg, Vol. II, pp. 189–206.

Richter, J.A., and G.F.N. Cox, (1984). "A Preliminary Examination of the Effect of Structure on the Compressive Strength of Ice Samples from Multi-Year Pressure Ridges," Proceedings Third International Offshore Mechanics and Arctic Engineering Symposium, ASME, New Orleans, pp. 140–144.

Sackinger, W.M., W.J. Stringer, and H. Serson, (1984). ''Arctic Ice Island and Sea Ice Movements and Mechanical Properties,'' Second Quarterly Report to U.S. Department of Energy, Geophysical Institute, University of Alaska, Fairbanks.

Weeks, W.F., and W.M. Sackinger, (1981). ''The Physical Environment of Arctic Alaska,'' in *U.S. Arctic Oil and Gas,* Working Paper No. 21, National Petroleum Council, Washington, D.C., 137 pp.

Voelker, R.P., F.W. DeBord, F.A. Geisel, J.L. Coburn, K.E. Dane, (1981). ''Winter 1981 Trafficability Tests of the USCGC Polar Sea, Vol. II, Environmental Data,'' Report to Maritime Administration, Arctec, Inc., Columbia, Md.

A Canadian Perspective on Arctic Offshore Operations and Policy

K. R. Croasdale
K. R. Croasdale & Associates, Ltd.

Much of Canada's Arctic is composed of seas and channels. In fact the earliest application of marine technology in Canada's Arctic (at least by Europeans) was by penetration of wooden ships. These ventures were conducted either for whaling or for exploration (such as the search for the Northwest Passage). Technology at that time was not really up to the challenge; as you know, many expeditions spent several years at a time in the Arctic and many of the explorers perished. However, it was during that time that much of the mapping of the Arctic took place.

Today, the major incentive to Arctic operations is the exploitation of minerals, primarily oil and gas. During the past 15 years or so, the Canadian oil industry, encouraged by government incentives and the prospects of large discoveries, has been very active exploring Canada's Arctic and offshore. Although no production has yet occurred in Canada's Arctic, several promising discoveries have been made. Gas and some oil in the Beaufort; gas and some oil in the High Arctic; gas off Labrador; and oil off Newfoundland. Furthermore, there is the promise of large oil and gas reserves yet undiscovered. Government sources predict potential recoverable reserves of oil for the Canadian Beaufort to be about 8.5 billion barrels; for the Arctic Islands 4.3 billion barrels; and for the Canadian East Coast to be about 12 billion barrels. Predictions for gas are also rather large amounts.

Of course, Canada also has large reserves of oil in the tar sands of Alberta. Frontier oil will need to be as cheap as the oil from the tar sands, for it to be developed. We are not sure yet whether this is going to be the case. However, the present attitude of the Canadian Government and industry is that both frontier and oil sands oil, need to be developed, in order to make up for dwindling production from the existing Western Canada oil fields. This perspective probably explains the Canadian Government's policy to implement very attractive incentives to industry to conduct frontier exploration. Under the present Petroleum Incentives Program up to 80 percent of the cost of drilling an Arctic well is paid for by the Government.

Now, we all know the major difficulty in exploring for, and extracting oil from the offshore Arctic is the presence of ice. In the Beaufort Sea, ice is present for about nine months of the year. Even in the summer, drifting ice creates difficulty for traditional offshore construction, dredging, and drilling.

MULTI-YEAR ICE

In the winter the ice cover over most of Canada's Arctic is continuous, and can contain features such as multi-year ridges. Shipping and offshore platforms have to be designed for interaction with such kinds of ice.

Off Canada's East Coast there is the additional problem of icebergs. Even in predominantly open-water areas, such as the Grand Banks off Newfoundland, icebergs create difficulties for offshore operations. Dealing with icebergs can be considered to be an Arctic engineering issue.

BEAUFORT DRILLING CONCEPTS

Moving back to the Arctic the Canadian Beaufort was one of the first areas to be considered for oil and gas exploration. In fact, permits were issued for exploration in the 1960s. So perhaps it would be appropriate to look at what has happened in the Canadian Beaufort since that time.

Potential hydrocarbons have to be drilled into in order to confirm their presence. One of the first tasks for the oil industry in the Beaufort was to consider the various methods of offshore drilling in ice-covered areas. Some of the different concepts being considered in the early 1970s range from artificial islands in shallow water to the use of drillships in the deeper waters in the open-water periods, ice platforms and steel or concrete gravity platforms.

Of all the concepts considered, the dredged island has been one of the most successful, especially in shallow Arctic waters. Islands have proven very adequate in resisting ice forces. But their cost effectiveness decreases with increasing water depth. In water depths beyond about 20m the fill requirements tend to be prohibitive, especially for islands with very shallow slopes. Also, islands are susceptible to wave erosion during the summer months, especially during construction. For these reasons, retained islands were conceived to both reduce fill volumes and provide immediate wave protection at the waterline. Most of the Canadian islands have been built with dredges operating in the summer. In contrast, most islands off Alaska have been built using land sources of granular material. To date, over 20 islands have been built and drilled from in the Beaufort.

ESSO Canada built the first island in 1972 and also, has built most islands in the Arctic in about 20m of water. One of the first designs for a caisson-retained island was ESSO Canada's. This consisted of a ring of steel caissons retaining an interior sand-fill. A different design of caisson-retained island was used at Tarsiut in about 22m of water. Here, four concrete caissons were placed on a berm and filled with sand to provide the retaining structure for an interior sand-fill. The Tarsiut island set a record for drilling at the very edge of the land fast ice in the winter of 1981–82. It proved well able to resist the ice forces, but had a rather limited surface area.

To help overcome storage problems, a self-contained mobile caisson design was developed by Gulf Canada. This caisson design has the rig permanently mounted, so that time for deployment and rig-up is much reduced. The Gulf Molligpale was built in Japan and is now at its first location in the Beaufort Sea, in about 25m of water.

A similar philosophy of a ready-mounted rig was adopted with the design of the Dome Single Steel Drilling Caisson (SSDC). This concept was designed for rapid deployment by water ballasting down onto a pre-built underwater berm at 9m. The main structure of the SSDC is actually the converted forebody of a large tanker, specially strengthened to resist the ice loads. It was designed and brought into the Beaufort Sea in less than nine months. The SSDC also holds the record for the deepest water depth drilled by a fixed platform in the Arctic. This was at Uvilule in 31m of water—well outside the land fast ice.

RESEARCH AND DEVELOPMENT

To support island and offshore platform technology development, a considerable amount of research was undertaken by Canadian industry and encouraged by the Canadian Government. Much of this research focused on ice forces, ice conditions and geotechnical issues. But it also considered oil spill clean-up technology and environmental impact.

To avoid duplication of effort and encourage cooperative research programs, the Arctic Petroleum Association Operators (APOA) was formed by industry in 1970. Since that time over 200 research programs have been conducted, and APOA has therefore been an important factor in the shaping of Canada's Arctic policy.

DRILLSHIPS

In the even deeper Beaufort waters, Dome Petroleum has pioneered the use of drillships which were brought into the Beaufort in 1976. The drillships are, of course, ice strengthened and by managing the ice with icebreakers, drilling can continue into freeze-up. However, it is not usually possible to drill much beyond late November, or when the ice is about 1 foot thick.

Drillships can only be used for a short summer season—hardly long enough to drill and test one well. For most of the year the drillships remain unused in a winter harbor. To extend the floating drilling season, the Gulf subsidiary BeauDrill designed and built a round drillship, the Kulluk. The Kulluk is designed to stay on location with 7 feet of ice moving against it. It is held on location with 12, 3½ inch mooring lines. So far Kulluk has drilled two discovery wells, and apparently has a performance which exceeds expectations. Incidentally, the Kulluk has recently been approved by the Regulatory Agency (COGLA) as a relief well drilling system for the fixed caisson systems described earlier.

ICEBREAKING SUPPLY VESSELS

To assist floating drilling and construction operations in the Beaufort, Canadian industry has pioneered the use of icebreaking supply vessels.

The most revolutionary was perhaps the Dome/Canmar Kigoriak; this is an experimental class-4 icebreaker and the first to operate year-round in the Beaufort (in a very limited fashion). The bow is spoon-shaped to reduce crushing and encourage bending failure. Also in the bow the ship incorporates a water spray system to reduce ice friction. The twin diesels supply 16,000HP through a gearbox to a single shaft and screw, which is shrouded to protect it from the ice.

HIGH ARCTIC ACTIVITIES

As already mentioned by Dr. Rey, the drilling in the High Arctic of landfast ice in deep water from ice platforms has been pioneered by Pan Arctic. Several gas and oil discoveries have been made. The Cisco oilfield in the High Arctic is likely to provide first oil production with some 300 million barrels of reserves currently indicated.

A possible subsea production scheme exists; but production is not likely before 1995, and will be very dependent on a viable transportation system.

CANADA'S EAST COST ACTIVITIES

Drilling off Canada's East Coast, using dynamically positioned drillships has been underway for years.

It has been possible to tow icebergs away from drillships; or as a last resort, to move the drillship off location before a collison. Two significant oil finds have been made off Newfoundland.

TRANSPORTATION

An issue not yet resolved is transportation. For oil this would be either by tanker through the Northwest Passage to eastern markets, or by pipeline down the Mackenzie Valley.

A large diameter gas pipeline, you will recall, was ruled out in 1977. The Berger Commission advised the Canadian Cabinet that construction of a large diameter line would have an adverse effect on local communities in the Mackenzie Valley. However, there is currently a small twelve-inch oil pipeline being built to Norman Wells from Alberta. By analogy it is considered that a small pipeline could be built to the Beaufort Sea by the early 1990s.

The alternative would be icebreaking tankers, some of which have already been designed by Dome Petroleum. These vessels, it is claimed, would be much safer than conventional large tankers because of strong double hulls, twin screws and rudders. Similar designs have also been developed for the shipment of liquified natural gas from Melville Island. Such liquified natural gas (LNG) tankers would follow the routes shown here. Currently the Arctic Pilot Project to ship LNG from the High Arctic has been shelved primarily because of the lack of markets for gas, rather than concern about the technical feasibility of the operation.

PRODUCTION CONCEPTS

To date, there has been no production from the Beaufort, but many different production scenarios have been developed. It is generally considered that initial production will occur from the shallow offshore using an extrapolation of the island technology already discussed. Although in the deeper waters fixed steel or concrete platforms may be more cost-effective.

A fixed platform will have to be designed to withstand impact by large but rare icebergs, and the designs for such platforms are currently being developed. An alternative to a fixed platform will be to use a floating system with subsea well templates. If an iceberg threatens the floating structure, then production would be stopped and a collision averted either by towing the berg or moving the platform. Such schemes are favored for early production because of lower capital cost than for that of a fixed platform.

SUMMARY

This has been a very superficial overview of Canadian operations and activities in its ice-covered Arctic waters during the past two decades.

Dr. Spilhaus emphasized that Arctic policy without action is not very useful. It seems to me that Canada's Arctic policy has encouraged action. Two major points of policy can be summarized as follows:

• Canada's Arctic policy has provided financial incentives to explorers for Arctic oil and gas.

• There is in place an Environmental Assessment Review Process which reviews plans for all major Arctic operations. This review process focuses primarily on environmental and socio-economic issues, and usually involves public discussion and hearings. These hearings, however, are not conducted using cross-examination by lawyers; the process tries to avoid the adversary approach.

To date the Environmental Assessment Review Process has looked at:

— Potential Beaufort Sea oil production (approved in principle)

— Davis Strait Drilling (approved)

— Lancaster Sound Drilling (not approved)

With respect to oil and gas production, the National Energy Board ultimately reviews and recommends to the cabinet whether a particular project is in the national interest.

In general, I personally believe that Canada's Arctic policy does encourage the goal of the development of mineral resources, and of the Arctic region. As far as I can judge, the policy and review mechanisms already in place should be very adequate in serving Canada's needs into the 21st century. I must emphasize that although I was replacing Dr. Roots as speaker on this topic, the views expressed are my own, and do not represent official Canadian policy.

THE U.S. APPROACH TO ARCTIC POLICY

Judith T. Kildow
Massachusetts Institute of Technology

I welcome the opportunity to speak to you today about the U.S. approach to Arctic Policy. Unlike several others on the program, I do not represent my government. Rather, I am an academician who has spent some time studying Arctic policy. Thus what follows are my own opinions on the U.S. approach to Arctic policies, and where it is lacking.

As a preview, in order to clarify the direction of my remarks, I shall be addressing the following areas: 1) the U.S. definition of the Arctic; 2) the U.S. national policy objectives for the Arctic as they are generally expressed in policy documents and statements; 3) the U.S. approach to Arctic policymaking as viewed through activities and interests which affect it; 4) considerations of time, place and events, which I don't believe are adequately being considered by the U.S. government; and 5) a summary and some of my conclusions about these things.

DEFINITION OF THE ARCTIC

Tucker Scully, the Department of State official in charge of Arctic and Antarctic affairs, speaking at an MIT conference in 1983, stated that the U.S. attempts to balance the physical properties of the Arctic with the human uses of the Arctic in defining the scope of policy-making. He placed the physical location of the Arctic between 60 and 65 degrees north latitude. This includes part of Alaska, but not the panhandle, the Aleutian Islands and Bering Sea, the Yukon Territory, the Northwest Territories, the Northern half of Quebec, Labrador, Greenland, Iceland, Svalbard, the northern fringes of Norway, Sweden, Finland, and European Russia, and large areas of North and East Siberia. (See Map 1).

The human activities included within this geophysical definition which represent the areas for policy concern, or what may be termed the functional definition for the Arctic, include the following:

- national sovereignty and jurisdictional claims
- national defense
- gas, oil and other mineral developments
- transportation and navigation
- oceanic and atmospheric research
- living resources and environmental management
- native interests and self determination.

Yet another way to define the Arctic for policy purposes is to categorize policy areas coterminous with nation-state interests. The breakdown looks like this: those nations which actually touch the Arctic Ocean basin—the U.S., Canada, the Soviet Union, Greenland, and Norway. However, if environmental and climatic conditions are an important condition for interests, the circle can be broadened to include those nations that share those conditions in common: Finland, Sweden and Iceland essentially share

Map 1. Arctic Region.

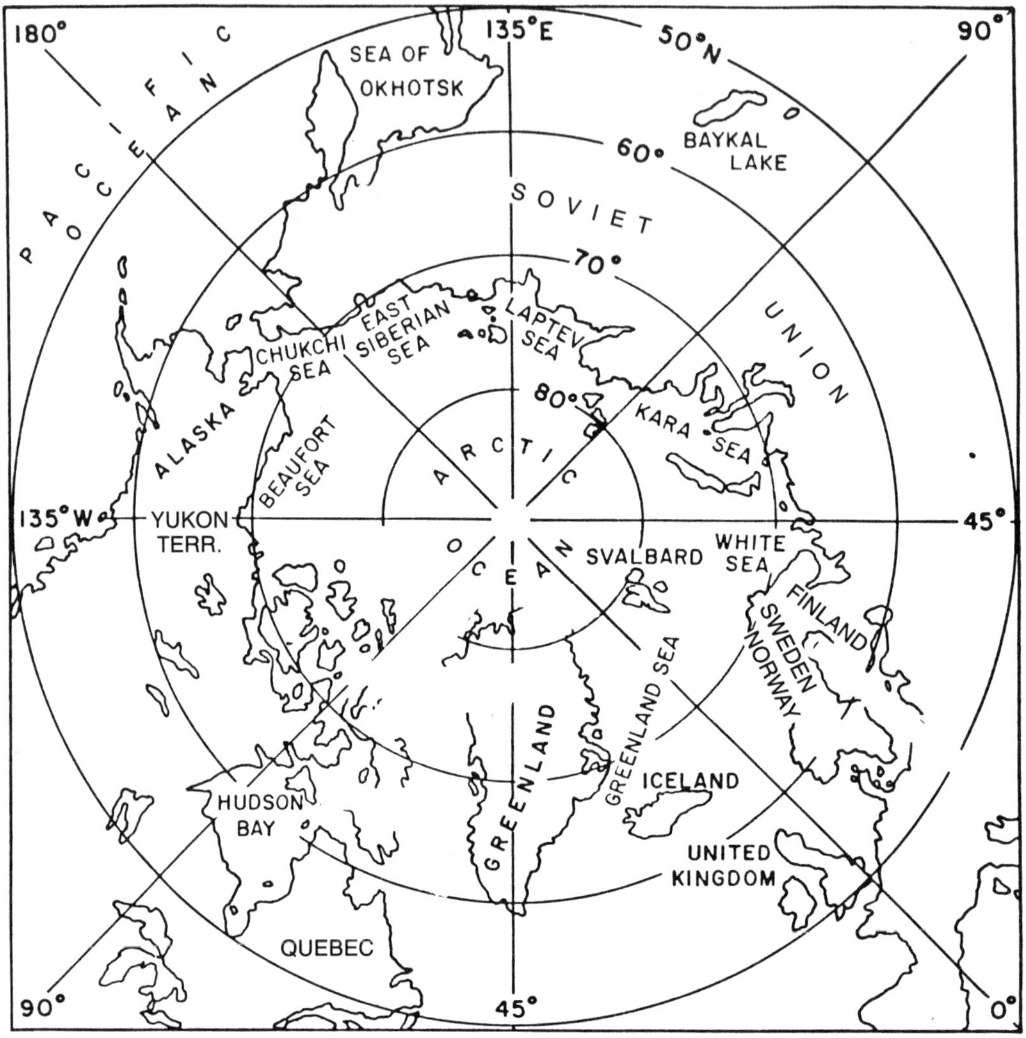

Source: Dyer, Ira and C. Chryssostomidis (eds.), *Arctic Technology and Policy*, Proceedings of the Second Annual MIT Sea Grant College Program Lecture and Seminar. New York: Hemisphere Publishing Corp., 1984.

the same climatic conditions. Taking this analysis one step further to include those nations which have definite strategic, political or economic interests in the Arctic, the list would expand to include the United Kingdom, France, West Germany, and Japan. This last set of definitions is particularly important if a positive approach to the Arctic is the objective, since identifying common denominators can assist in finding areas for cooperation.

U.S. POLICY OBJECTIVES

The national policy objectives for the United States have been expressed at least three times in the past fifteen years, beginning with a very general, noncontroversial, and not very informative memo in 1971 and ending with a far more specific piece of legislation in 1984. In 1971, the *National Security Decision Memo #144* stated the following four objectives:

1. support sound and rational development in the Arctic
2. minimize adverse environmental effects
3. promote mutually beneficial international cooperation
4. protect essential security interests.

These general statements were indicative of a nonspecific policy and little concern for the Arctic within U.S. policy priorities at the time. The Arctic stayed on the back burner for quite awhile, except for a growing U.S. military interest in the area (not to mention the private corporate interests of the petroleum industry). Much more government interest went toward the Antarctic during this period, in large part due to the renegotiation of the treaty for that area.

Finally, in 1983, the Interagency Arctic Group, the same entity responsible for the 1971 memo, was revived and it released another memo in June of that year. This memo, *National Security Decision Directive #90*, had many of the same platitudes of the first, but added to it were two new items, or at least old items put in a new way:

1. renewed and strengthened emphasis on resource development in the Arctic
2. a minimum of federal presence in the Arctic to accompany these new developments.

The first policy put environmental interests ill at ease and the second dismayed the federal agencies that ordinarily have activities and interests there—the Coast Guard, whose search and rescue and oil spill clean up mandates seemed compromised; and, the National Oceanic and Atmospheric Administration and the Environmental Protection Agency whose tasks of carrying out research and monitoring environmental impact seemed threatened. While received poorly by many interests, the U.S. policy of a small federal civilian presence in the Arctic was consistent with the prevailing policies of the current administration to diminish government's role.

A strong scientific program of international cooperation survived this memo and continues to flourish, albeit, mostly supported by the U.S. Department of Defense program carried out by the Navy. The National Science Foundation continued its Polar research programs for the Arctic, but on a very small scale.

The third pronouncement of U.S. interests and objectives in the Arctic is the *Arctic Research and Policy Act of 1984*. The purposes of the Act as stated therein are:

1. "to establish national policy, priorities, and goals and to provide a federal program plan for basic and applied scientific research with respect to the Arctic, including natural resources and materials, physical, biological and health sciences, and social and behavioral sciences;

2. to establish an Arctic Research Commission to promote Arctic research and to recommend Arctic research policy;

3. to designate the National Science Foundation as the lead agency responsible for implementing Arctic research policy; and,

4. to establish an Interagency Arctic Research Policy Committee to develop a national Arctic research policy and a five-year plan to implement that policy.'' (See Part IV, Appendix B).

The findings enumerated in this legislation reveal that Congress had concluded, much as I had, that the U.S. has paid too little attention to the Arctic, and as a result has not established a rational set of research priorities based on a rational understanding of the interests and needs of the United States in the Arctic. The findings indicated a recognition that other nations had passed us by in their capabilities to use and study the Arctic and that it was time the U.S. did something about it. A program of research priorities was needed and this legislation was to ensure that it was established and carried out.

While this legislation should be somewhat assuring to Americans—that finally their government is recognizing the importance of this area and is going to do something about it, it is disconcerting to observe that no funds have been appropriated to carry out the mandates of this Act, but only a statement that necessary funds should be appropriated later under the normal rules for appropriations, not a terribly reassuring statement in view of current budget-cutting trends.

A specific definition of the geographic boundaries of the Arctic was given in this legislation. ''Arctic means all United States and foreign territory north of the Arctic Circle and all United States territory north and west of the boundary formed by the Porcupine, Yukon and Kuskokwim Rivers; all contiguous seas, including the Arctic Ocean and the Beaufort, Bering and Chukchi Seas; and the Aleutian chain.'' (See Section 112 of the *Arctic Research and Policy Act of 1984*).

U.S. APPROACH TO ARCTIC POLICY

The U.S. has not had a consistent approach to policymaking for the Arctic as I shall attempt to demonstrate. In fact, I would express it as a policy of vacillation. First, there has been a schizophrenic attitude about how to treat the Arctic—whether as a distinct region or as a group of separate, functional activities and interests. The U.S. Government claims to handle it as a region, yet none of the policies outlined above would indicate a firm regional approach. Understandably, it is difficult to perceive it as a region from a political point a view, since it is comprised of such a diverse set of nations with diverse interests. Indeed, it is difficult to see this area as a homogeneous area where all interests are similar and parallel. To have regional policies for the Arctic is very difficult, except in very limited ways, as some of you pointed out today. Perhaps environmental protection, or environmental monitoring might be one area for a regional policy, but it is not really benign as you know, with the immense interests in development focusing on that area.

The Arctic has also been connected, at least within the institutional structure and infrastructure of the government, with the Antarctic, and we are all aware of the tremendous policy differences between the two areas. While each enjoys the same climate and many of the same scientific issues and problems, they are very different politically, economically, and physically. To link them together, I think, has given the Arctic too much of a back seat; as a result, the U.S. has concentrated mostly on Antarctic policies to the detriment of U.S. understanding of the Arctic. While the U.S. has focused its efforts toward the South Pole, the Canadians, Soviets, and others have surpassed the U.S. in many areas of understanding and working in the Arctic. It would be far easier, for example, to place the Arctic in the same category with the Mediterranean, an enclosed sea, than to put it with another polar area, which has no political and few economic similarities, for purposes of policy understanding and policymaking.

Another area of vacillation of the U.S. approach resides in the changing emphasis from coastal zone problems to free transit problems. Our vacillation can be attributed to the events of the moment, but seems to have little other rationale, particularly any rationale linking these two interests.

U.S. desire for international cooperative research efforts appears strong and consistent. This is understandable in view of the fact that the U.S. has limited capability to carry out some of the research the military needs to have done. This positive, consistent approach to Arctic research is overshadowed by concern about the Soviet presence as well as about the lack of Soviet involvement in scientific cooperative ventures.

There is, however, room for vacillation here as well as in the next six or seven years. The U.S. has called for a strong research approach to the Arctic in its new legislation which could have two implications. The first, and positive one, could be that the U.S. will continue its same path of international cooperation, but contribute even more expertise to the efforts, especially if the work is carried out through the National Science Foundation, named as the lead agency to set the priorities and programs. However, if the NSF plays only an advisory and facilitating role, which is all it is asked to do in the legislation, then the Navy could feasibly continue its leadership in the research program, as it continues to today. With its own capabilities it may feel less need for international cooperative efforts and more need for proprietary work, excluding possibly even civilian U.S. scientists, depending upon the NSF role, Congress, and the attitude of the administration in power at the time the capabilities come on line. I consider this a serious possibility and one to be avoided.

Another problem which has produced an uneven approach by the U.S. Government is the degree of presence it has had in the Arctic. The vacillation between a strong presence and a hands-off attitude in the Arctic has depended on the prevailing philosophies of the administration in power and on bureaucratic momentum, depending on which aspects of national interests are being considered and which programs. For example, the U.S. Government has a very strong presence through the National Marine Fisheries Service for some of the native Americans, in a rather paternalistic sense. It is only recently that the Congress has tried to set up a system to get the U.S. Government out of that situation. It is not a simple problem to resolve and may take many years, so that federal presence may dwindle but not disappear on that particular issue for some time.

On the other hand, the government has recently taken a hands-off approach to resource developments in the Arctic, leaving the petroleum industry alone, except for a few precariously funded environmental research programs carried out by the U.S. Departments of Interior and Commerce. Here, the vacillation between presence or no presence is serious and complex. On the one hand, the Arctic is a place where the government's presence is necessary because of conflicts and potential conflicts. For the public's interests to be served, the federal government needs to have a presence.

Military security, environmental impact, and economic security of native Americans are a few of the reasons for continued presence. Yet, the current policies regarding presence are not consistent with these needs, not to speak of the many other areas not mentioned here.

My final perception of the U.S. approach is who sets the policies for the Arctic. Here again the vacillation is apparent. Verbally, the civilian sector in the name of the Department of State and the NSF is credited with making our Arctic policies. In fact, most, if not all of the policies appear to be dictated by the Department of Defense and the Navy, where most of the money is spent and most of the expertise existing in this country resides. As stated before, it is not clear whether or not the *Arctic Research and Policy Act of 1984* will change this trend.

CONSIDERATION OF TIME, PLACE AND EVENTS

The vacillating nature of U.S. Arctic policies over the past fifteen years indicates that there has not been a serious appraisal of the problems and opportunities presented by the Arctic. Some of the issues I have mentioned will continue to become more serious and draw us into deeper conflict; others will grow less serious and perhaps even disappear, either having been overtaken by events or technology, or actually having been resolved due to simultaneity of certain conditions.

Hence, it is important to assess U.S. national interests in terms of the current and projected concerns, using a time framework. If we try to project, for example, when some of the problems may peak and determine which may peak together during the same time frame, we can at least plan a bit better. This approach to analysis gives a basis for demanding that the government give a detailed analysis of each of the areas of concern—those listed under functional definition—to determine the degree of seriousness that simultaneous peaking of certain problems may have. Implicit here is the assumption that correlations among activities and interests can be identified and synergisms in the form of conflicts and opportunities, can be predicted.

On the other hand, the U.S. may be overly concerned about some issue or set of issues, which over time will decline in importance, and it would be essential to know where to place priorities. For example, thinking in terms of a time frame, we know there is going to be accelerated activity, and hence, probably more technological advances. We will have easier access to the resources. With more military operations in the Arctic, the probability of military confrontation and jurisdictional disputes could increase. Using a horizontal axis to represent time and a vertical axis to represent a deepening of concern, correlations begin to appear as one looks ahead. This might be a useful tool for negotiating certain issues in the near future. (See Figure 1).

SUMMARY OF OBSERVATIONS AND CONCLUSIONS

In conclusion, I would like to summarize the observations I have made here today. While U.S. policies have vacillated for the past fifteen years, they appear to reflect current U.S. interests in the Arctic. Military interests appear to dominate, understandably since the Arctic is a place where the borders of the U.S. and the Soviet Union come the closest, and therefore represent a potential area for conflict between the two

Figure 1. A Time Framework of U.S. National Interests.

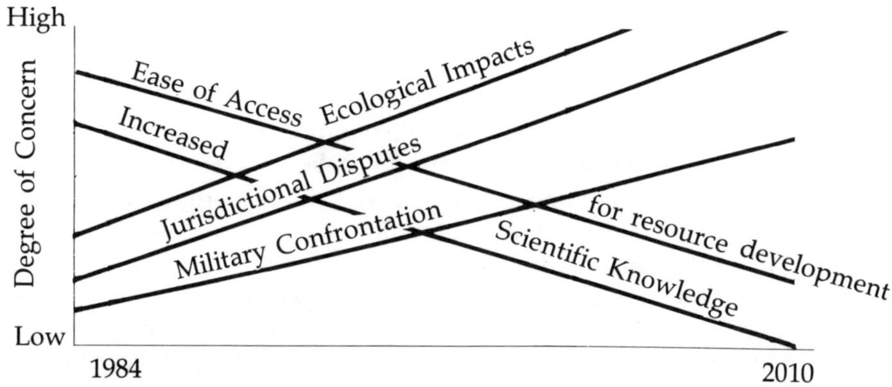

Assumptions: growing nationalism, increased resource, investments less cooperation.

superpowers. Scientific interests seem to be emerging strongly, at least since Congress has expressed its interest through the *Arctic Research and Policy Act of 1984*. Private oil and gas interests are being left pretty much to themselves, consistent with current administration policies of less government presence in all things but military. Those private interests are presently operating at a low level due to market considerations, and thus will arouse little public interest until those activities accelerate again.

The vacillation of U.S. policies with regard to the various national interest areas poses a serious problem for carrying out a rational Arctic policy, particularly until officials are unable to see the complex correlations among these interests and activities. Perhaps the new legislation will force more consistency, although as I pointed out, it is not easy. National policies could inhibit resource development just because jurisdictional disputes could prevent resources from being exploited. In fact, just this summer, the United States leased lands beyond what some nations believed was its jurisdiction. The result of this lease sale will be interesting to follow. The U.S. federal presence could help resolve some of the jurisdictional conflicts if, in fact, the government decides to have more of a presence in the Arctic. The absence of a federal presence could create a vacuum for the powerful, and therefore public interest might not be served. There is obviously going to be increased activity in the realms of resource developments, transportation, and military and scientific research. And with this increased activity comes increased potential for conflict, again, indicating the need for federal government attention, presence.

Finally, increased opportunities for new international cooperation are created by all these increased activities. Efforts towards cooperation could lead to a limited regionalism, a promising future. If we could find more ways for collaborating and cooperating, perhaps the Arctic could become a form of a region. If the nations with interests in common got together by means of a treaty or a pact, it would be helpful and could reduce the potential for serious conflict. On the other hand, there are so many areas for conflict, and nationalism appears as strong as ever, that the positive scenario seems remote.

Discussion

MR. PAGE: I notice, Dr. Kildow, in your list of agencies that have influence on Arctic Policy, the glaring absence of two entities: One is the United States Congress, and the other is the State of Alaska.

DR. KILDOW: You are quite correct. The new legislation passed by Congress certainly encourages a more serious focus on Arctic Policy. It thus far has not set any policy or influenced any. I was concentrating on current policy. It wasn't an oversight.

DR. SACKINGER: Could you comment on the likelihood of future Arctic cooperation in the form of bilateral versus multilateral treaties or arrangements?

DR. KILDOW: Bilateral arrangements are always easier to create, but I must say there is an interesting precedent that has been set in Arctic research where numerous nations have been collaborating for some time. In this case, there might be an exception in that Arctic nations have been cooperating in international scientific research all along. It might be that the continuation and expansion of that could be very possible, although, as I pointed out, the Soviet Union has not been an active participant. I also must refer to my perception of the possibility that in time the United States may also shy away from such cooperative efforts.

DR. GERWICK: I read in the paper that the world court has just settled the boundary dispute between Canada and the United States on the East Coast. Do you think that this forms a sort of pattern that might help to settle, let's say, eventually the line on the West and Beaufort Sea and then, possibly, be extended to some of the other nations where there is a boundary dispute?

DR. KILDOW: American fishermen are obviously dissatisfied with the decision and I am not sure how the United States will respond to them. The West Coast poses a different set of conditions, however, and can not be compared, except to say that the United States, may, in its wisdom, decide to settle boundary disputes to get on with resource developments.

ADM. VENZKE: I have a question about the Department of Defense having the major influence on policy in the Arctic. I know there are a lot of military interests in that general area; however, as a result of a recent polar icebreaker requirement study, I talked to a lot of people, and I don't see where the Department of Defense is having an overwhelming influence at all. Could somebody, or could you, give me an example of where that is happening?

DR. KILDOW: First of all, anybody who tries to look at American policy won't find many statements of it, as I indicated. I think one of the reasons you don't find overt public pronouncements or policies is precisely because the Department of Defense, which shrouds many of its activities in secrecy, does dominate Arctic policy. It shares funds and programs with other agencies, but retains the power of the dollar and decisions over them. That has been my perception. I watched the struggle go on with the interagency memo in 1983. That memo went round and round for quite awhile in the federal government due to disagreement among agencies. But really, the lead was taken very strongly by the Navy, and while the memo allegedly came out of the State Department, the Navy had a large part in it.

ADM. VENZKE: From my perception, though, I can't come to that same conclusion. Maybe I have to do a little homework.

DR. KILDOW: As I said, your perspective is going to be different from mine; I look at it quite differently.

MS. BALES: In response to the Admiral's comments, I might add that perhaps you based your assessment of the Department of Defense's role on the amount of dollars the Navy has put into that. It is substantial.

DR. KILDOW: Well, there is more to it than that. However, when you put in dollars, you learn things and have the expertise; if you look at the amount of money that any agency in the government, outside of DOD, is spending in the Arctic, you will find very little.

MR. PERKINS: First I would like to comment that our observations in studies we are just completing at the Marine Corps National Research Council would tend to say that the Minerals Management Service has a pretty strong economic policy in the Arctic.

DR. KILDOW: I would say a strong economic interest, yes, in fact, the policy of hands-off is not unappealing to that office. The Department of Interior role is going to be very dependent on the oil market, although Alaska is thought to be one of the richest areas. However, it may be reasonable to expect that economics will drive part of the policy and the Department of Interior will continue to vie for a strong voice to assure OCS developments there.

MR. PERKINS: The second question is that I thought we did have a bilateral agreement with the Soviet Union on the sharing on Arctic technology, and I wondered whatever happened to this.

DR. KILDOW: As I said, the Soviets have not been very cooperative.

DR. SPILHAUS: What is your advice and conclusion to those of us who want to act in the Arctic? Simply because there is a possibility of increased conflict, do we back off?

DR. KILDOW: Growing nationalism is not going to help matters at all. We really have to strive in those areas for all possible cooperation. We must keep after our government to maintain an open, cooperative attitude.

DR. SPILHAUS: So your conclusion is that we should have cooperation between those nations concerned with the Arctic?

DR. KILDOW: I would definitely think so, but I think it may be difficult to convince our government and others to follow that path.

PART III:

Workshop Reports

ARCTIC WORK SYSTEMS

Leader: Lt. Cdr. Lawson W. Brigham
U.S. Coast Guard

Discussion in the Arctic Work Systems workshop emphasized forecasting for the years 2000 to 2010, while keeping in mind the traditionally long development times required for the Arctic. The discussions gave consideration to future needs, challenges to technology, constraints (e.g., geographical, political and economic issues) and scientific needs for each topic. The topics discussed included:

- power systems
- navigation systems
- oil spill response/recovery
- driving, ROV's and other submersibles
- search and rescue.

POWER SYSTEMS

The discussion on power systems centered on two alternatives to systems based on the Carnot cycle: nuclear and fuel cells. Nuclear power systems were viewed with some skepticism due to thermal and radiation pollution, cost and geopolitical considerations. It was the sense of the Workshop Group that small nuclear plants (as power sources) could not be buried in the Arctic seabed due to the political sensitivity of such an action. Fuel cells offer promise as low maintenance, untended solid state power sources. One drawback to current fuel cells is that they require sizable quantities of fuel and oxidizers to produce relatively small amounts of power. However, technological advancements are expected and fuel cells have recently had satisfactory testing aboard the M/V *Empire State*. More development is needed prior to the year 2000 if fuel cells are to become completely effective and viable for Arctic applications.

NAVIGATION SYSTEMS

Advanced navigation systems are needed for both surface and subsurface navigation throughout the Arctic basin. For surface navigation the Workshop Group noted very poor LORAN coverage and expressed doubts about the availability of Global Positioning System (GPS) information. Subsurface navigation is complicated by the obvious fact that surface snow and ice interfere with the satisfactory penetration of radio waves. Options for under ice navigation include both active and passive acoustic devices that would be placed on the seabed. Active devices would appear to be easier to use than passive systems, but they would require more servicing. Positioning systems for acoustic undersea devices must be developed. Shallow water acoustic problems must also be addressed before development of an effective, subsurface navigation system.

OIL SPILL RECOVERY/RESPONSE

Problems and challenges in oil spill response and recovery were divided into seven problem areas: detection, surveillance, containment, recovery, storage, transfer, disposal and dispersants.

Detection and surveillance have primarily been conducted by visual means in the Arctic. Ice cover, and broken ice and snow make visual detection difficult, if not impossible, especially for large scale surveillance. Synthetic aperature radar from aircraft is now being tested in the Arctic and has shown promising results for ice detection. Other possible solutions to this challenge are suggested such as airborne or satellite synthetic aperature radar, lasers and submarine-carried upward-looking acoustic devices.

Containment and recovery challenges are centered on the ability to separate oil from ice rubble. Booms, pumps and transfer systems presently in operation have not proven effective in cold environments. The ice that flows near the boom skirts draws oil with it, and the pumps for pulling the oil from a containment area choke on the ice fragments.

Transfer and disposal are complicated by the same problems that beset any transportation in the Arctic. The best option discussed would appear to be *in situ* burning. Dispersants are a possible solution to many oil pollution problems outside the high latitudes. The fact that dispersants generally work poorly in cold water and that they require a good deal of mixing energy makes their use appear to be limited in the Arctic. However, there has been little research done to confirm or deny dispersant capabilities in this environment. The use of ice as a dispersant has been suggested, but no study has been done on the effectiveness of such an approach. The politically sensitive nature of oil spills may make this approach unattractive.

DIVING AND SUBMERSIBLES

The underwater environment in the Arctic presents few challenges not already met by diving, ROV and other submersible operations presently being conducted around the world. The problems for underwater work in the Arctic occur on the surface and at the water/ice/air interfaces. Some of the problems for divers could be solved by using self-contained modular habitats or submarines to house divers or to transport them between work sites. Divers are slowly being replaced by ROV's and manned submersibles, especially for routine cleaning and inspections. Submersibles are affected similarly as divers at the air/ice/water interfaces. Potential solutions by the year 2000 appear to be untethered remote vehicles and high endurance submersibles. The major challenges to the development of these vehicles are in the command, control and communications systems for operations under sea ice. Background noise levels may be high enough in the Arctic to pose significant problems for acoustic communications.

SEARCH AND RESCUE

Search and rescue (SAR) in the Arctic presents special problems due to the long distances, isolation and extreme environmental conditions. The first problem in search and rescue is to determine that an emergency situation exists. One method of determining this is to use satellites to monitor emergency frequencies and to require all parties working in or traveling through the Arctic basin to carry EPIRB's (emergency position indicator radio beacons) or ELT's (emergency locator transmitters). This type of detection has proven itself highly effective in various parts of the world. It was the sense of the Workshop Group that, from a regulatory point of view, all parties moving in the Arctic should be required to file "float or flight plans."

Search and rescue can be accomplished from a variety of platforms, each having inherent advantages and drawbacks. Fixed wing aircraft are effective search platforms because of their long range and ability to receive/transmit large volumes of communications. Most also have the capability to carry heavy detection equipment. However, they usually cannot land on short, unprepared runways which is the normal situation for most remote regions of the Arctic. The employment of vertical takeoff aircraft in the Arctic may be one future solution. Although helicopters can land almost anywhere and have proven effective while deployed aboard many of the world's icebreakers, they are severely restricted by range and weather. Surface craft such as icebreakers are too slow to be used as a SAR response vehicle. Air cushion vehicles (ACUs) show promise because of their speed and range, but they have yet to prove themselves reliable enough for SAR work. Because none of these solutions is totally adequate, parties that work in or transit the Arctic basin must be self sufficient during the time interval between the actual emergency and rescue. For offshore structures, the ability of personnel awaiting rescue to be self sufficient should be the primary concern of all operators. Contingency plans must include adequate survival systems and describe locations where personnel can establish ice camps to await rescue.

WORK SYSTEMS AND ARCTIC MARINE DEVELOPMENT

Forecasting the work systems required in the Arctic by 2000 proved to be difficult. The development of various support systems is inexorably linked to the progression of offshore operations around the Arctic basin. Figure 1 depicts one potential sequence for development in Alaskan waters. No consensus was reached by the Workshop attendees as to which stage of development portrayed might take place by 2000. Many felt that operations would primarily be conducted in the marginal areas of the Beaufort and Chukchi Seas. The transportation mode would be icebreaking surface vessels and support systems would develop with shallow water as a constraint. Others believed that development might well be in the deep offshore zone where submarine tankers could conceivably operate. There would be requirements to develop acoustic seabed devices for submarine navigation, deep seabed power systems, deep diving systems and enhanced rescue systems for the polar pack environment. There is little doubt as to the diverse challenges posed to designers of work systems for each of the ice environments listed in Figure 1. However, the timing of offshore development will have a significant bearing on the types of overall systems needed by 2000.

WORKSHOP KEY ISSUES

At the conclusion of the workshop each attendee was asked to provide a list of key issues affecting the future of Arctic work systems. The emphasis of these key issues was placed on three important categories:

1. the assessment of technological need

2. regulatory requirements
3. management approaches.

The following is a summary of key issues that should be addressed prior to the year 2000:

Technological Need

- development of ROV's and submersibles (to improve manipulative ability and under ice navigation and positioning; to develop means for dealing with the air/ice/water interface)
- air cushion vehicle rescue capabilities
- continued development of vertical lift aircraft for use in the high latitudes
- cold environment firefighting problems
- ice penetration technology
- surface support for ROV's, submersibles and diving
- satellite technology for (close gap) navigation below the ice, search and rescue (locator systems), ice thickness measurement and under ice oil detection

Regulatory Requirements

- regulations to prevent unprepared parties from venturing into the High Arctic
- safety standards and backup systems for ice operations
- enhanced international cooperation with regard to Arctic oil spills
- mandatory emergency locator devices

Management Approaches

- *Arctic Research and Policy Act of 1984:* a golden opportunity to provide guidance and research funding
- innovative search and rescue guidelines (e.g., recompression chamber locations, federal/private storage, general SAR guidelines for the Arctic)
- international policy for Arctic information and data
- assurance of the availability of Global Positioning System (GPS) information for all Arctic users

Figure 1. Expected Sequence of Alaskan Offshore Development.

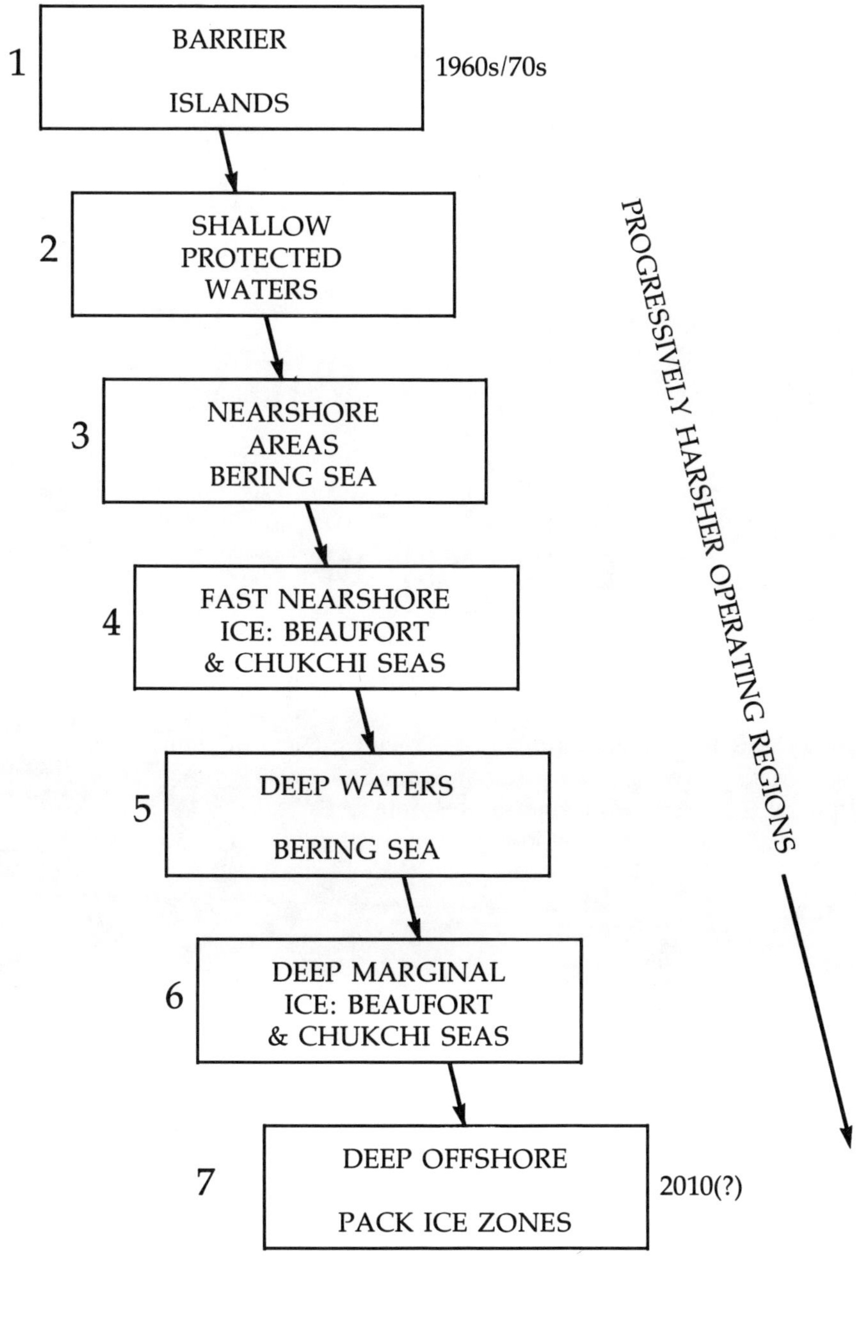

ARCTIC OIL AND GAS (A)

Leader: Ernest Pallister
Pallister Resource Management, Ltd.

In reviewing completed offshore structure platforms, one sees a trend toward concrete and light-weight concrete structures and another trend toward hybrid (concrete/steel) structures. The appropriate design is largely a function of the Arctic Zone in which it will be located. The zones being: Sub-Arctic (1st year landfast ice), Multi-Year Ice Zone, and the Iceberg Zone. The designs that seem appropriate for the various zones are:

Sub-Arctic Zone: semi-submersibles, jack-ups, subsurface rigs and fixed-tripod structures
Multi-Year Zone: gravity cells, combination gravity cell/leg
Iceberg Zone: combination gravity/leg structure.

Future designs must utilize sub-sea facilities and passive and active ice management systems.

Ocean Engineers are faced with significant challenges for the Arctic region. The extensive detailed engineering problems require the best technologies. The economics of oil production in the Arctic has a very significant impact on Arctic natural resource development.

The Arctic environment varies and ice forces cannot be considered by themselves. Other environmental concerns include winds, currents, sea state and biological impacts. Based on various environmental factors a probabilistic approach must be used.

Past experience and current technologies indicate the following structures appear to be appropriate:

Shallow Water — gravel islands
Mid-Depth — combination concrete/steel structures
Deep-Water — combination gravity/pole structures.

Caution must be used since no one solution is correct. One must select the best design for the particular location. Attention should also be focused beyond exploration to production and transportation. Two feasible methods of transportation are icebreaking tankers and pipelines.

Considerable time has been spent on discussions of production platforms. Without significant 'finds,' future production may be limited. In the United States, past "lease/sale" agreements have involved water depths less than 60 feet, thus enabling the use of gravel islands. Lease/sales now extend to water depths of 400 feet with 80 percent of the area having a depth less than 200 feet. Depths greater than 60 feet will require production structures to be concrete/steel cones, with deeper regions using a combination of gravity and floating structures. Caution must be exercised with some "gravity structures" due to some locally poor soil conditions.

The obvious concern of Arctic offshore platforms is ice loading. Ice creates two fundamental problems: large lateral forces and a very short open-water season for construction. When predicting ice forces/stresses, it is important to consider the limits of ice forces/stresses. This is categorized into: *Limit Stress* (infinite ice driving force), *Limit*

Momentum (iceberg impact), and *Limit Force* (large multi-year floes). Further consideration should be focused on winter vs. summer ice. It appears that multi-year pack ice governs for summer ice designs for depths up to 100 meters. For winter ice, land fast ice governs design for depths less than 25 meters and pack ice governs for depths of 25-100 meters.

Data are becoming available from 25 sites (depth 31m) in the Canadian Beaufort. The interesting finding is that actual ice forces are lower than anticipated. Actual forces are believed lower due to: fracturing processes, large-scale strengths being less than laboratory studies, and a worse combination of events being rare. The data are based on first-year ice.

Data are also being obtained for impact loading due to multi-year pack ice at Hans Island, Canada/Greenland.

ENVIRONMENTAL CONSIDERATIONS

Leader: William Sackinger
University of Alaska

Co-leader: M. Brewer
U.S. Geological Survey

Discussion initially centered around the concept that good engineering normally implies a low failure probability, and thus a low probability for environmental damage. By analogy with exploration activities on land, it was recognized that seismic exploration, exploratory drilling, transportation, and disposal of debris should be accomplished without leaving permanent traces offshore. Production sites with lifetimes of 40–50 years were seen to be possible. It was felt that public awareness and concern for the quality of the environment would continue over that 50-year period.

Definitions of the Arctic were extended to include regions of cold temperatures, rather than a simple latitude definition. The effect of operations upon humans, as part of the environment, was also seen to be of major importance.

There was discussion of the seriousness of oil spills, and the higher probability of smaller, incremental oil spills. An oil spill "countermeasures brigade" for the Arctic, with optimized equipment and strategy, was suggested. The role of industry and of the Coast Guard was discussed.

Environmental diagnosis and subsequent understanding of naturally occurring relationships was seen as a major task in the 21st century. The use of numerous stations to collect data in an automated manner, using projected advances in computer electronics and communications techniques, was felt to be important. In the case of field trips by scientists, interdisciplinary teams should be deployed to gather a comprehensive and contemporaneous data set. Logistics support should be provided to take such data at remote Arctic locations.

Results of such research should be relayed to policy makers and regulating authorities in clear terms to facilitate decisions on offshore operations by industry. The organization and funding of Arctic research within the federal government was perceived to be in need of coordination and integration, and the recently passed Arctic Research Policy Act was expected to aid in that respect. Multi-year continuity in research programs, in institutional and professional commitment to Arctic problems, and in Arctic information storage and retrieval, was seen as very important. A national commitment to build an Arctic research and engineering cadre was suggested.

RECOMMENDATIONS:

1. Arctic institutions for science and engineering are needed to educate new people in the Arctic specialties; continuity over a long time frame is needed.
2. Research capabilities, facilities, and scientists must be developed, thus encouraging entire careers in Arctic research.
3. Methods and funding should be provided for logistics support of Arctic science at

a variety of locations in the Arctic, which will not be fixed and which will be remote.

4. Environmental, engineering, and statistical information about the Arctic should be made available in summary form for government policy makers, regulators, and research managers.
5. A central institution for Arctic science and engineering information collection and dissemination was recommended, which could also serve as a center for visiting scholars on Arctic subjects.
6. A need to resolve the issue of the merit of restricting offshore drilling in the U.S. Beaufort during the productive summer drilling months in order not to interfere with the mating of bowhead whales.

TRANSPORTATION

Leader: George Vance
Mobil Oil Canada, Ltd.

This session addresses the transportation needs in the Arctic and Subarctic in the early 21st century. Basically we shall address:

- what the needs will be in the years 2000 to 2010
- what technology will be required to meet these needs

It is anticipated that future needs will be similar to current needs but more demanding in levels of intensity and reliability. They will involve transportation of people and goods for:

- strategic reasons
- exploration of resources
- construction of facilities
- operation and maintenance of facilities
- rescue and escape
- raw material and hydrocarbon transport.

These needs will have to be met with the refinement and development of the following:

- surface ship technology
- above-surface transportation technology such as ACV's and STOL aircraft
- subsurface technology such as submarine tankers, remote controlled submarines and pipelines.

The following presentations will address these needs and the technology that is being developed to meet them.

THE FUEL CELL PROPELLED SUBMARINE TANKER

William Kumm, Arctic Enterprises

The validity of the submarine tanker is based on the availability of a deep ocean route from the Alachan North Slope and the Canadian Archipelago to a deep water ice-free port in Norway. The scenario is based on converting Arctic gas to methanol and moving it with six 165,000 DWT submarine tankers. The submarine tanker would have a non-nuclear, fuel cell powered propulsion system using methanol and liquid oxygen. The fuel cell is very efficient in that it converts chemical energy directly to electrical energy. No special technology is required; it is presently available but remains to be utilized correctly. In addition the submarine tanker described permits 94 percent of the weight of the tanker to be used for cargo transport.

Details of the design and operational requirements and costs of the system are contained in a seven-volume report prepared for the United States Maritime Administration which is available through the National Technical Information System.

Discussion

Q. Why is the tanker a square rather than a body of revolution?

A. It is more economical to build a rectangular shape than a round shape and the 13-15 knot top speed requirement does not require the body of revolution shape.
Q. Is the ballast mixed?
A. No, cargo and ballast are separated.
Q. What is the crushing depth?
A. The tanker can be designed to operate down to 1000 feet or greater, however, there was no need to go below 700 feet, which was the design depth.
Q. Would there be vertical stability problems with a submarine 800-900 feet long?
A. No. Studies on large submarines indicate dive or climb attitudes would not vary more than $\pm 5°$.
Q. Is there a significant hazard in carrying liquid oxygen and hydrocarbons?
A. Yes. New regulations and new technology will have to be developed to ensure that risks are minimized.

SUBMARINE TANKER PERSPECTIVE

Larry Jacobsen, General Dynamics

Most of the submarine tanker design work is now in the open literature. General Dynamics has been working on submarine tanker design since the mid-1950s, however, we feel that a large 150,000 DWT tanker will not be required in Beaufort Sea until a large amount of hydrocarbon transportation is required and thus may not be required until the middle of the next century. We do feel that a 40,000 DWT tanker is a viable cost-competitive system for shuttlement of oil and gas between the Canadian Arctic Islands. We feel the technology is available to design and build cost-effective submarine tankers. However, there are areas which require further development:

- crew escape and rescue
- submerged cargo transfer
- submerged docking
- operational control, a particular for a nuclear powered vessel.

Discussion

Q. Is there a cost estimate for such vessels?
A. Not for the year 2000. Cost estimates have been made and it is our feeling that a submarine tanker system will be competitive with a pipeline or surface tanker under certain conditions.
Q. Who will support research and development in this area?
A. We don't know. General Dynamics has done considerable development work in this area but will not be putting too much more into it. I suppose industry will when the need arises in the future. Or, a non-profit arctic research institute should be formed, supported by government, industry, and academia to assure this type of research is accomplished on a continuing basis.

MODULAR SUMARINE TANKER

Jack Chappuis, Vevey Engineering Works, Ltd.

We have looked into the design of a modular submarine tanker where the cargo-carrying portion of the submarine is disconnected from the propulsion and control section of the submarine. The system is viable for shuttling oil over short distances (a few hundred miles or so). The large module can be left at the site for storage, if required.

SOVIET ARCTIC MARINE TRANSPORTATION SYSTEM

Lt. Cdr. Lawson Brigham, USCG

The Soviets have been leaders in the development of Arctic transportation systems. The following tables will illustrate their accomplishments, the extent and age of their icebreaker fleet and the matrix of their operational requirements. It seems evident that the Soviet fleet will require replacement in the next twenty years and a forecast of the possible replacement program is contained in the following tables. It appears the Soviet trend will be more and more toward nuclear icebreakers.

Table 1. Soviet Arctic Marine Transportation System: Accomplishments.

- world's first experimental ice tank for ship research (1955)
- first nuclear powered icebreaker (1959)
- development of the world's most powerful icebreakers
- first transit to the North Pole by a surface vessel (1977)
- first icebreak escort of a freighter across the Arctic Basin (1978)
- routine operations of icebreaking cargo vessels (SA-15 class)
- employment of specialized icebreakers (shallow-draft polar & river icebreakers)
- substantial ice piloting expertise due to extensive, long-term efforts
- achievement of a nearly year-round navigation season on the northern sea route
- successful application of new technology (spoon-shaped bows, ice breaking-ice removing devices, cp propellers, bubblers, quadruple-screw propulsion systems)

Table 2. Average Ages of the Soviet Polar Icebreaker Classes at 2000–2010.

		Average age	
Class	# in class	2000	2010
Bresnev/Siber	2	24	34
New Nuclears	2	14	24
Lenin	1	41	51
Ermak	3	25	35
Moskva	5	31 youngest	41
Kapitan Dranitsyn	2	20	30
Kapitan Sorokin	2	23	33

Table 3. Feasibility Matrix for the Soviet Arctic by 2000.

Mode of Marine Transport

SEAS	submarine	icebreaker escort	independent cargo ship ops
Barents	YES	YES not economical	YES
Kara	NO shallow/terminals	YES	PROBABLE
Laptev/ East Siberian	NO undev	YES	NO ice/undev

Table 4. Forecast for 2000.

— POLAR 8 and fleet block obsolescence will force building 150,000 SHP ship
— fewer, but larger tankers and cargo ships (on independent ops)
— routine application of nuclear to shallow-draft polar icebreakers
— couple nuclear and LASH concepts for transport vessels
— await western development of submarine tankers
— new channel clearing devices
— expanded use of aircraft aboard large ships
— improved integrated tug-barge operations
— development of the operations infrastructure

Discussion

Q. Do the Soviets have tugs in addition to icebreakers?

A. Yes, but they don't serve the same purpose as the larger icebreakers. There are about eight tugs.

Q. If the Soviets are building so many nuclear icebreakers, will they allow them to be built in another country such as Finland?

A. Most likely they will have the hulls built in Finland with the nuclear plants installed in Russia.

Q. Why do the Soviets require so many icebreakers?

A. The basic philosophy is different. The Soviets have a specific plan carrying them out to 1995. They have established year-round settlements in the Arctic that require finished cargo transported in, and raw material exported out, eleven to twelve months a year. Marine transportation in the Arctic is a routine undertaking and does not generate the excitement that is associated with Arctic transportation in the United States and Canada.

ARCHIMEDEAN SCREW TRACTOR (AST)

M. Oshima, Mitsui Engineering and Shipbuilding

The AST is an amphibious tractor of a new type, operated in sea ice for transportation of personnel and materials, as well as for various works such as ice-breaking, oil deployment, etc. It can break through ice of below designed thickness, and climb up from in-water onto ice when ice is thicker than designed and continue running.

The Archimedean Screw Vessel (ASV) is a highly efficient icebreaking vessel suited to shallow water. Four screws, attached to the vessel's bow around the water line, interact with ice by means of rotating blades of screws. These pull the vessel's bow onto the ice, breaking the ice by the vessel's weight. In the 21st century, ASV's will be useful for supply logistics, pipe-laying, dredging, foundation soil arrangement, anchor handling, oil spill recovery, etc.

Discussion

Q. Are there any unique applications anticipated?

A. Yes, we are looking at putting the screws on a bow and attaching it to a tourist vessel in Japan.

Q. Have you done any testing?

A. Yes. We have tested a 20-meter model of the larger vessel with the screws at the bow in the ice north of Okido, Japan. With the screws at the bow, the vessel is more effective than a conventional icebreaker because the screws help it climb up on the ice and apply more of its weight in breaking the ice.

Q. Can the vessel ram with the screws at the bow?

A. No.

Q. Can the vessel be dropped 60-80 feet onto the ice?

A. No, not as it is currently designed.

Q. What is the horsepower used?

A. 400HP forward and 150HP aft. They are separately controlled with separate prime movers.

ARCTIC AIR CUSHION TECHNOLOGY

Ero Maksiueu—Wartsila Arctic, Inc.

ACV's have been in general development for twenty-five years or so but have only been considered for the Arctic in the last few years. They will not replace surface ships or helicopters, but they will have a role in certain operational situations. Originally, we anticipated two roles for the ACV, one icebreaking, the other as a transportation vehicle over ice. Although the ACV had icebreaking capabilities of its own, it will not replace the surface icebreaker. However, as a transporter, it can be used to support operations year round. It can move cargo from ship to shore and it can be used as transportation between communities.

From the tests we have conducted, we feel an ACV arctic transportation system is feasible but the ACV cannot be designed using aircraft technology. We decided to use marine technology—i.e., welding in lieu of riveting, diesel engines in lieu of turbines, etc.

In North America there may be one or two ACV's used occasionally in the Arctic. In Finland we have one experimental craft, in Russia they have forty in operation transporting cargo from ship to shore. Wartsila has just delivered fifteen non-self propelled ACV's to Russia for use in transporting cargo and materials. We anticipate a 100-DWT design in the near future and 200-DWT in the next 20 years, however, it will be a long time before a 300-500 DWT version is built.

Wartsila has concentrated on an Arctic ACV, having a medium speed, i.e., up to 30 knots, a simple skirt configuration, using welded construction and higher speed diesels. The vehicles have a low foot sprint pressure in the area of 1 to 5 PSI and do not do any damage to the tundra. They are designed to go over steep ridges about the height of the skirt. If the ridges have a shallow slope the ACV can negotiate it.

In closing, at Wartsila we feel the ACV will definitely find a place in Arctic transportation as the need arises—as it has in the Soviet Union.

ARCTIC OIL AND GAS (B)

Leader: Jerry L. Machemehl
ARCO Oil and Gas Company

INTRODUCTION

This session focused on the problems confronting design engineers, in the present and in the future, with specific emphasis on ice structure foundation interaction problems. Participants were asked to identify their highest priority for Arctic development in the 21st century. The response came mainly from consultants who felt the oil industry needs to improve its record in Arctic development. Specifically, to:

- target costs more effectively
- improve knowledge of ice strengths
- redevelop *Bulletin RN* to adjust for deficiencies in the design curves and equations
- fund more work in deep water soils
- be more aware of the problems of trying to place foundations on Arctic silts in areas of shallow gas.

Discussion also raised the following needs:

- greater mobility during the winter
- permission to drill in the summer season
- increased reliance on marine seismic exploration.

ICEBERG RESISTANT OFFSHORE STRUCTURES

L.C. Zaleski-Zamenhof, C.G. Doris

The development for the 21st century needs to be planned in conjunction with the design philosophy and technology of materials. Progress must be from a deterministic (past) to a semi-probabilistic (present) and, ultimately, to a probabilitstic (future) design.

Structures collapse and zero risk is associated with infinite cost. There is a probability of failure of 10^{-5}, i.e., for every 100,000 structures, an average of one will fail per year. Actual failure probabilities in some structures range from 10^{-10}, to 10^{-8}.

- it is necessary to define the accepted probabilities of failure and progressive collapse
- it appears important to develop a balanced design for all environmental events
- in ice-infested water, the choice between a floating platform and a bottom-founded one should be based on water depth
- limited sliding is acceptable as it contributes to energy absorption
- in the future, Arctic offshore engineering will need to use a second level probabilistic approach, it appears the best way to proceed is by Monte Carlo simulation.

GENERAL DISCUSSION

- on the capacity to disconnect floaters before a major event
- on concentrating moorings in certain directions to protect against icebergs

- on designing floaters for impact
- ridge problems might best be handled with a ridge at the edge of the floe
- problems of research funding (private vs. university)
- high cost of investigation forces selection of the lowest bidder, thereby compromising quality
- problem of disseminating information.

OBJECTIVES AND CONCLUSIONS

The general consensus of workshop participants was for:

- a need to target costs
- an agreement on a unique set of data
- increased marine seismic profiling in ice
- permission to drill in summer months
- deep ocean submarine drilling
- low-cost methods for marine construction (e.g. ferrocement)
- a need to have greater oil industry representation at future symposia.

Scientific Research

Leader: Ira Dyer
Massachusetts Institute of Technology

Scientific research is presently underway on a large number of issues thought to be important in development and use of the Arctic Ocean. Principal among these research issues are:

- sea ice mechanics
- electromagnetic signatures of sea ice for remote sensing
- physical oceanography
- meteorology
- marine soil mechanics.

Such research is likely to continue not only because of its intrinsic interest to the community, but also because of its potential in support of technological development and exploitation of the Arctic Ocean.

Perhaps the most important thing about the present course of scientific research relates to the transfer of knowledge between the two broad communities with a stake in this effort—the research community and the user community. This transfer seems at best sporadic and incomplete. Workshop participants sensed this via questions from researchers directed at users on scientific research needs and, in contrast, from users to researchers on knowledge acquired to date. Symptoms of the transfer problem are easy to find:

- disparate funding agencies for researchers and users
- largely different professional groups.

Some Workshop participants suggested remedies, such as establishment of an Arctic Research and Engineering Institute in the United States, but as Chairman, I ruled that such institutional recommendations had best be left for consideration to other groups at other times. It is my personal view, however, that a conference such as Arctic Engineering for the 21st Century is an important step in the transfer process. But it is a first step, and MTS along with other professional groups should face the challenge of evolving continuing transfer opportunities. It is also my view that transfer works best when communicators can be in frequent face-to-face contact. How MTS or others can best encourage this, I leave for future deliberation.

In looking to the 21st Century, Workshop participants identified several goals which seem worthy of pursuit, and which would serve to focus scientific research beyond the thrusts already in place. The more important of these goals are:

- ice route planning
- capacity of the Arctic for accommodating wastes
- *in situ* conversion of coal and/or ocean thermal energy
- use of permafrost layers for sub-seabed structural implantments
- artificial polynyas for focused bioproductivity.

Some additional comments on the first two are given here, largely to reflect the sweep of interests expressed by the Workshop participants.

Satellite remote sensing technology along with *in situ* data buoys and advances in basic sea ice mechanics offer the promise of identifying sea ice conditions most suited to surface transportation. As a function of time, regions of converging versus diverging ice conditions can be predicted which in turn relate to surface ship operations. As an important scientific aside, uncertainty exists on the severity of ice conditions that might be expected in the 21st century. One school of thought has us facing thicker and rougher ice of greater latitudinal extent. Another, based on the CO_2 greenhouse effect, predicts substantial shrinking of the icecover in the next one hundred years, perhaps enough by the time of the 22nd century to keep the Arctic continental shelves open through the winter. Presumably, this scientific controversy will be resolved before we enter the 21st century. But either way, we believe it both useful and possible to bring scientific knowledge to bear in support of surface ship operations, via route selection.

Many questions remain with respect to the fate and impact of pollutants in the Arctic. Solid, liquid, and gaseous wastes, including both continuous and pulsed discharges, should be considered. Scientific research should take as a goal the quantitative understanding of the Arctic's capacity to bear these wastes. In almost all cases we will be faced not with zero environmental change, but with finite changes that need to be traded off with quantitative exploitation benefits.

All such goals which serve to focus scientific research, and the pursuit of Arctic science for its intrinsic values, depend upon work processes or technologies for their ultimate success. Some of the more important technologies identified by the Workshop participants are:

- use of satellite-based systems needs to be increased dramatically for acquisition of synoptic data
- a research submarine capability should be added to more standard platform capabilities such as research icebreakers, offshore rigs, drifting ice islands, drifting ships, and underwater manned habitats
- research should be integrated more highly than heretofore, so that we can understand as a whole the system of thought identified with physical, chemical, biological, and geological oceanography; atmospheric science; and ice science
- research should include increased concern for effects localized in space and/or in time since these, often more so than large-scale averages, connect with concerns such as environmental insult, threat to structures, and ease of travel.

I close with a compliment to the organizers of this Symposium. I have enjoyed and profited by my attendance. I sense that the Workshop participants, through their many ideas and questions, also think the experience worthwhile.

Arctic Minerals

Leader: George Gryc
U.S. Geological Survey

The Arctic Minerals Workshop reviewed the geologic framework of the region north of the Arctic Circle in Alaska and discussed the mineral (exclusive of petroleum) potential of the region. Only two mineral commodities are currently in production: precious minerals, gold and silver; and construction materials, sand and gravel. However, recent discoveries (since the late 1950s), to a large extent spinoffs from petroleum geologic studies, have delineated a belt of world-class copper deposits and prospects in the southern Brooks Range and zinc-lead-silver deposits in the central and northern Brooks Range. It is predicted that both will be in production in the 21st century and will require marine transportation to markets from terminals and offshore loading platforms on the northwest coast of Alaska. A plan to mine and concentrate zinc, lead and silver ore at the Red Dog Prospect has been published and an Environmental Impact Statement has been prepared. The plan [will] includes a 200-person camp, a road (56-73 miles), a port facility with about a two-mile causeway and an offshore loading platform. This project is expected to be in production by the 1990s. Production from other prospects of both zinc-lead-silver and copper will follow this development and continue well into the 21st century.

Vast coal resources have been estimated for the coastal plain region of the North Slope. Published estimates range from 100 billion up to 4 trillion short tons of low-sulphur coal. The coal rank ranges from lignite A to high volatile A bituminous. Tests indicate coals of coke-blending quality. It is predicted that as world supplies of oil peak and diminish, these coals will be utilized, certainly by the mid-21st century. Some will be mined and shipped from port facilities along the northwest coast, but some will be utilized *in situ* by conversion to other energy modes and transported in that form. Geologic extrapolation indicates that these coals extend offshore into the Chukchi Sea. Use of these resources offshore by some *in situ* techniques is likely perhaps by the end of the 21st century.

Perhaps the most critical mineral commodities to the development of other resources in the Arctic are the construction materials, sand and gravel. It is a little appreciated fact that a virtually inexhaustible supply of gravel overlies the Prudhoe Bay petroleum deposits. Elsewhere on the North Slope, gravel is scarce and west of the Colville River, almost totally absent. Pea-size beach gravels occur along the coast in some local areas, but most of the surficial deposits in the region are silts and sands. Thus, should a major deposit of petroleum be discovered in this region, there would be a significant engineering problem in terms of utilizing these finer materials in constructing roads, work pads, etc. Extensive research in the use of these materials will be required and may be a critical problem in the 21st century.

Other than coal, sand and gravel, no other mineral commodities are known or can be forecast in the offshore of Arctic Alaska. No one in the workshop could document offshore mineral commodity occurrences, although there is some evidence for metallic module mining off the coast of Siberia. Geologic environments in the spreading center of the Arctic Basin may be favorable for hydrothermal deposition of cobalt, copper, nickel and other base and rare elements. There is a need to study the High Arctic Basin and to evaluate the mineral potential.

The workshop pointed out the need for more basic knowledge and the exchange and interchange of this knowledge in several basic sciences. The mineral resource potential of the Alaskan Arctic and the Arctic Basin in general is very high. However, more research is needed to provide the basis for assessment of the impact on other resources, including wilderness, wildlife, native subsistence, etc., and in turn, to assess the risk of unacceptable impact on these other resources. These research data are also needed for application in engineering design and operational procedures.

The workshop applauded the promulgation of the Arctic Research Policy Act and other such institutionalized efforts that will aid and support research and disseminate the results for all to use. A regional emphasis on the Arctic was considered at least as important as that currently attributed to the Antarctic. Adequate funding of the Arctic Science Policy Act is viewed as essential to the wise management and development of Arctic resources.

Sea Ice Management

Leader: Warren Denner
Science Applications International Corporation

I was responsible for chairing the Symposium session on sea ice management, and Dr. Ira Dyer acted as the recorder for that session. We had some 50 people in the session, and had a variety of inputs on this broad subject area. Figure 1 gives you an idea of a comprehensive view of sea ice management. It is not a simple task that has to be carried out by an individual on a drilling platform or the captain on a ship transiting the pack, but the whole process of decision making that continues from the time that one contemplates operating in a particular region until the time that activity is successfully completed.

The ice-covered seas of the Arctic, as we have heard from others at this Symposium, contain very large reserves of oil and gas and other minerals. The safe development of these resources along with other aspects of the northern development represent one of the significant engineering challenges of the 21st century. There doesn't seem to be any argument about that.

Careful management will be required to meet this challenge, however, in the sensitive Arctic environment; those of us who have lived and worked in the Arctic for a period of time know how sensitive it is.

Figure 1. Elements of a sea ice management system.

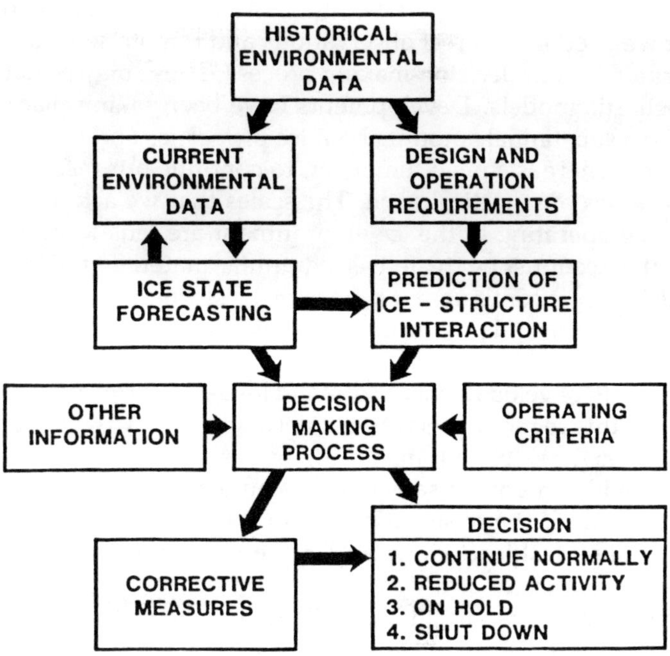

Successful sea ice management will be a very important aspect of this development, crucial to a number of activities. I have taken a very broad view of sea ice management, to which no one in our session found objection. *Sea ice management is a comprehensive understanding of ice and ice processes which will allow for the safe and successful conduct of activities in cold regions.* There are a tremendous number of activities that go on related to any kind of development, and therefore the problem of management is multifaceted. We have to look at different aspects of management for fixed structures, surface transportation, subsurface transportation, aircraft operations, and surface effect vehicles.

Drilling and production are certainly activities that are going on and will continue. Coastal facilities will have to be built. Dredging operations will be necessary to construct offshore platforms and to lay subsea pipelines. We have flooded sea ice to build platforms in the past and will probably do so in the future—maybe with some more gigantic schemes than we have seen so far. We cannot overlook the responsibilities for controlling pollution which certainly implies some form of ice management. According to the report from the session on Arctic Minerals, hard mineral development may take place in the near future, which will bring in other considerations than those associated with the offshore oil and gas development. We can not overlook the military operations that are going to go on, and certainly the military will be considering many of the questions that we are confronted with in industry ice management programs. We must also consider the living resources with their relationship to the ice environment and therefore management in that regard will be required as well.

The management tools that are available to us are our knowledge of the ice properties and processes. Data bases of one sort or another—and I do not just mean ice but those aspects of the ocean and atmosphere which force the ice conditions—must be obtained over a substantial period of time so that we understand the variability.

Observations are of crucial importance and are becoming easier to obtain with finer resolution, but we need both direct observations and remote sensed data. Models are extremely important to our decision-making process. These may be deterministic, conceptual, or stochastic models. Developments have been taking place in these areas. We have learned a substantial amount about ice properties and processes over the last couple of decades. There is a need, however, to continue our research on sea ice properties and processes on a tactical scale. The scales that we actually have to contend with when we are operating in the ice environment are tens to hundreds of meters. We now know that ice must be treated as a granular material on these scales. It is not a homogeneous isotropic continuum. I believe that a stochastic formulation with respect to ice, its granular nature and properties, is a subject for further research between now and the year 2000.

There is a need to develop further short- and long-range ice and weather forecasting techniques and in this regard observational networks are extremely important. There are approximately eight satellite transmitting buoys in the Arctic at the present time. That does not provide a very dense network. You may be interested in knowing that only about 60 percent of the observations from those buoys get into the world data net in a timely enough fashion to have any influence on the analysis which is the first step in large-scale numerical atmospheric forecasting.

Atmospheric, oceanographic and ice-related data bases exist for the Arctic. However, there are some serious gaps in these data bases, particularly on tactical space and time scales. Data bases in the Western World consist very largely of data that have been collected by Western nations. There is a large body of data that exists in the Soviet Union which is not freely accessible to Western scientists. Sea ice data bases need to be expanded, refined and integrated. I don't know how many of you have tried to work with sea ice data bases in their existing format, but it is not a trivial matter.

A real time, all-weather, integrated ice monitoring system will be required for successful operation in the year 2000. The system should consist of satellite, airborne and surface-based integrated observational networks which provide an input to a hierarchy of models. The technologies and the models exist or are on the drawing boards, but the difficult problem is the implementation and integration of the systems into a decision-making process. Also, more research is required with respect to the remote sensing of sea ice, particularly the radiometric properties of ice and the classification of ice types. In that regard, the scientific community is at a low point at the present time with respect to remote sensing capability of ice from satellites.

Future satellite-based systems are uncertain, in terms of the sensors that will be flown and the orbits of the satellites. However, they represent an essential part of the system that will be required in the year 2000. We have seen recently the unexpected loss of satellite-passive microwave data during a period of time when it was being used operationally to support activities in ice-infested waters. The government responsibility to provide remote sensing and other observational data must be decided and implemented. We can argue forever about the responsibility for putting satellites in space and acquiring and utilizing the data. I think as long as it is unclear about the government's role in this area, and the availability of that data to the civilian sector, we are going to have problems. A national policy on this matter should be developed and implemented.

It is essential to devise remote sensing techniques to determine ice thickness and integrity which, in turn, allow for increased development of the transportation and off-shore industries. I note that the Soviets are reported to have an ice thickness sensing radar which can be flown in front of icebreakers or ships in transit.

More development is required to provide near real time ice warning systems for our offshore developments, particularly fixed structures. Electromagnetic and acoustical systems seem to show promise in this regard.

At this time we are unable to cope with even a small oil spill in ice-covered waters. The fate of the spilled oil is not well known, and more research is needed to develop techniques for dealing with oil spills in ice-covered waters.

It appears that the problems associated with the exploration of oil and gas in shallow water will be solved substantially by the year 2000—the technology is in hand. However, concern was expressed about exploration and production in deeper water, and that may remain the challenge for the 21st century. Transportation systems—and by that I not only mean tankers and pipelines, but the shore facilities associated with these—remain significant problems.

Living resources are strongly influenced by ice and ice processes. Everybody seems to recognize this fact. Many questions remain unanswered concerning the living resources of the ice-infested regions. Effective ice management will be required to protect and utilize these living resources as well as to carry out our activities in those waters.

More work is needed to identify and mitigate extreme conditions. We seem to be fairly comfortable dealing with what we consider known environmental conditions in

Arctic, but we are not confident that we have identified or can mitigate extreme conditions, such as very large icebergs, large multiyear floes or pressure ridges. Until we can feel comfortable in this regard, I am afraid we are going to be faced with over-engineering and accepting undefinable risks. To the extent that we can identify and mitigate extreme conditions, then we should be able to reduce costs and risks.

Engineering is a very important aspect of ice management, as are the techniques involved in operating in ice environments. If you want to tow an iceberg, there is a certain technique involved which requires some learning experience and some trial and error. Experience is important, and certainly planning is very important. In the final analysis another management tool is regulation.

The Soviets have, I think, developed transportation systems in the Arctic well beyond any other nation. That represents a very substantial technology base to them which should not be overlooked by the Western World. They have had a very large centralized program of development since the early part of this century. As we have heard, they have achieved nearly year-round shipping in parts of the Northern Sea Route, made effective use of SEV technology, explored for offshore oil, operated submarines under the pack ice, and developed the largest ice breaker fleet (including three nuclear powered ships) in the world.

Discussion

MR. GERWICK: Thank you, Warren. I see Dr. Assur in the audience. Dr. Assur, as I understand it, CRREL does translate a great many of the Russian documents, and don't you have a computer printout of what is available on the Russian scientific data publications?

DR. ASSUR: Yes, correct. We have covered literally what we can get hold of at CRREL. The computer printout is available to people who are interested. Also, by subject, you can push a button on a computer and get access at any university and so forth. Oil companies are doing it right now. I don't know the extent to which they are using it, but they do have access. We also have every item on microfiche so we can inspect it immediately which is very efficient. Within an hour or two we get the information we want back.

DR. DENNER: I appreciate what you say with respect to your bibliography which is, indeed, an excellent one. It was not long ago, however, that I visited your laboratory, and your librarian told me that CRREL no longer translates Soviet papers except on special request, and there were very few of those requests coming from your own scientists internally. How much of the literature is Soviet?

DR. ASSUR: About two-thirds of all literature in this subject area is in Russian.

DR. DENNER: A point I would like to make is that if you look over the literature that is published in the Western World, you will find that there are very few Russian citings in that literature. At the same time, if you review the Soviet literature you will find a substantial number of Western references cited.

ARCTIC MATERIALS

Leaders: Masaneo Oshima
Mitsui Engineering and Shipbuilding Co., Ltd.

Richard Seemel
Bechtel Civil and Minerals, Inc.

Mr. Oshima gave an overview of Arctic materials which included structural steel, concrete, hybrid structures, and geotechnical materials and foundations. In his overview he requested comments by the participants relating to preceding workshops where requirements for materials may have been cited. He also requested that participants look ahead 15-20 years to judge what may happen by the turn of the century.

Mr. Itoh of Nippon Steel Corporation spoke next, summarizing his paper entitled "The Trend of Development of Steels Used in the Arctic Ocean Field." He predicted that future developments in the steel manufacturing industry will spring from the present new "clean steels." These will permit dry welding at great sea water depths by remote automatic processes using electron beams, lasers, and other planned concentrations of high energy. He also predicted the development of new corrosion-protective coatings (including those for repairing) such as urethane, polyethylene and ceramics; improved defect detection systems; improved repair techniques; and concluded with a plea for better information exchange between the users of steels in the Arctic with the manufacturers of steels, so that the user's needs can be met in the future. The workshop concluded in discussions that such information feedback could happen through some sort of "Arctic Institute" sponsored by all concerned parties.

Dr. Fiorato of Construction Technology Laboratories, a Division of Portland Cement Association, spoke next about "Concrete for Arctic Construction." He spoke particularly about strength/weight ratios in concrete structural elements, new developments in high-strength lightweight concretes, durability of concrete in the Arctic, developments in admixtures for concrete, the need for protection of reinforcement in a saltwater environment, and coatings for concrete. He cited examples of high-strength, lightweight concrete used in Global Marine Development Inc.'s Concrete Island Drilling System (CIDS). This concrete had a fresh unit weight less than 115 pounds per cubic foot with actual test compressive strengths of more than 8500 PSI. The structure was built on a fast track schedule with excellent quality control. It is currently operating in the Arctic under lease to Exxon Co., U.S.A. Looking to the future he noted a continuing improvement in concrete technology, producing lightweight, durable concretes using, for example, silica fume, superplastizers, extended life superplastizers, coated lightweight aggregates, and coatings for reinforcing steel. He stated—and the workshop members agreed—that technology for considerable improvement in Arctic-compatible concrete is in hand, but must be translated into practice.

The workshop members brought up such considerations as the economy of the CIDS and the problems associated with fast tracking. The need for more *constructable* designs and the ability to *industralize* the construction process were noted.

There was also comment about possible new kinds of "concrete" such as resin matrix materials (later in the workshop the use of asphaltic concrete was discussed, as well).

Mr. Koski of ABAM Engineers, Inc., then spoke about research into steel-concrete hybrid structural members (composite systems) for "ice walls" around a drilling platform. These composite structures use steel membranes as permanent forms for concrete. The economy of such structures has excellent potential. Investigations to determine required design strengths of the composite are underway. Analytical methods are being developed to refine designs; it is believed that composites can be designed and built strong enough to resist iceberg impact as well as ice floe forces. Ease of repair of composite systems was also noted.

Factors affecting composites which are under development were discussed, including:

- use of plate membranes on both sides or on a single side
- tie throughs (shear connectors)
- arch-shapes to resist ice loading
- durability as affected by moisture content of the encased concrete
- prestressing.

Mr. Gerwick of Ben C. Gerwick, Inc., and the University of California, Berkeley, then briefly discussed the future outlook for enlarged versions of CIDS. He cited the advantages of being able to construct CIDS at different shipyards and then marry the pieces—either at a staging location or at the drilling location—after floating them into position. He reviewed different configurations of CIDS and the need for building the larger base platforms in place, as well as the construction of various component pieces. In addition, he made several suggestions for designs of steel-concrete hybrid structures for Arctic use.

Richard Seemel of Bechtel Civil and Minerals, Inc., then led a discussion of future developments in the use of geotechnical materials in Arctic regions for both structures and foundations. He cited several developments since 1960 which have affected construction in general. Under the heading of "Construction Equipment and Methods," he described improved cutter head dredges and tunnel boring machines (TBMs). He predicted further improvements such as sea-going dredges and improved TBM's that will influence future operations in the Arctic and the sub-Arctic.

He cited the improvement in gravel filter technology since the 1960 ties, the use of plastic filter fabrics which have already improved construction of earth-fill drilling islands, and the development of improved consolidation techniques for silts and other fine grained soils. These developments will have a future impact.

Workshop members commented about the use of asphaltic concrete, made from local beach sand, for slope protection on manmade islands; the need to design for ice loading rather than wave action (in some Arctic locations, the reverse is true); and the use of "sacrificial" materials by deliberately overbuilding and allowing deterioration of islands to take place, to cite some examples.

The possibility was also raised about the use of designed ice structures for Arctic construction. Another suggestion was the use of tunnels for access and for pipeline conduits to underground drilling stations (assuming TBM improvements result in reduced cost). These concepts were challenged from a safety standpoint in case of blowouts. The desirability from an environmental protection standpoint was noted.

The use of saltwater beach sand with a cementitious binder for construction was discussed at length. In the High Arctic, on the flat shoreline adjacent to the shallow continental shelf, rock is unavailable. Beach sands are presently unusable for aggregates because of the salt content which can pose a problem by traditional portland cement concrete standards. It was the belief of some engineers present that, in the 21st century, such difficulties will be overcome and we will see the increasing use of beach sand as aggregates for building structures and sea walls. Such use may require a series of steps to process fine particles into larger sizes, or a pelletizing process. Also, consideration of corrosion prevention of reinforcement steel is essential. It was also stated that, where bedrock lends itself to tunneling, drilling access, and storage, caverns will be developed using TBM's as their efficiency and usefulness improves. The present trend in TBM development, the progress to deeper and deeper underwater drilling sites, and environmental concerns dictate use of tunnels.

In summary the Workshop saw for the 21st century:

- improved steels for cold temperature use
- facilitated use of portland cement concrete as we catch up with present developments in high-strength concrete admixtures, coatings, and construction techniques
- increase in the use of hybrid structures
- development of new codes and standards for material for Arctic structures
- use of ice as an Arctic structure material
- use of other cementation components for ''concretes'' in addition to traditional portland cement, i.e. asphaltic materials or resins
- tunnels and caverns for access, drilling sites, pipeline conduits and storage as we progress into deeper water
- application of soil consolidation methods to foundations.

PART IV:

Closing Plenary Session

Oil Strategy and Ocean Law in the Arctic: A Scenario for the 21st Century

Willy Ostreng
Fridtjof Hansen Institute

We have not come to the northward limit of communal progress. There is no northern boundary beyond which productive enterprise cannot go until North meets North on the opposite shores of the Arctic Ocean as East has met West on the Pacific.''[1]

V. Stefansson, 1922

Stefansson's vision retains its sobriety. Nevertheless, there is a difference between past and present; at Stefansson's time most people regarded the Arctic as a region where there was nowhere to go and nothing to do when you got there; the potentials of the region were ''surplus to requirements.'' Today the industrialized world requires its surplus and knows what to do with it when being there. This difference is not yet dramatic, the conquest of the Arctic for military and economic purposes is still inchoate. Certain trends are emerging, however, allowing for a discussion of the directions developments might take before ''North meets North.''

This paper examines developments in the Arctic within the fields of military strategy and petroleum business. The purpose is to identify some of the major driving forces behind these developments and to assess their strength and duration in order to come to grips with some of the more likely consequences for permanent petroleum-related activities in the North. We shall also discuss the effects on the uncertain state of ocean law in the Arctic and conclude by examining alternative scenarios with respect to a regulated and orderly exploitation of northern resources.

WHAT IS HAPPENING WITHIN THE FIELD OF STRATEGY AND PETROLEUM EXPLORATION: A BRIEF SKETCH

The Militarization of the of the Arctic Ocean

The Soviet Northern Fleet—one of the world's most potent in firing power is based in the Barents and White Sea areas. Its outstanding characteristic is submarines; some 70 percent of the Soviet strategic submarines are stationed here. In comparison, the Northern Fleet disposes only 17 percent of the surface vessels.[2] This composition demonstrates that the overriding aim of this fleet is strategic deterrence.

Due to the development of short-range missiles during the 1960s, Soviet strategic submarines (Yankee-class) had to sail through the GIUK-gap between Greenland, Iceland and Great Britain in order to come within firing range of the American east coast. In 1972 the first Delta-class submarines became operative, and with a missile range of 8000km, targets in the United States, China and West Europe were within reach from a firing position in *Arctic waters.* Sailing into an adversary's waters was thus no more required. Today there are 24 Delta submarines in the North, constituting 75 percent of the Soviet Navy's vessles of this class; the remainder are stationed in the

Pacific. The most recent member of the strategic family is the Typhoon, which is twice as large as its predecessor. Only two have been deployed, but several more are on the bedding. This new generation of exceptionally large submarines has been the subject of much speculation in the West. Their purpose is probably found in their capability to operate in ice-covered waters. With 25,000 DWT and a 25m hull, the Typhoon can penetrate ice more than 2m thick. As the thickness of the ice averages 3 meters and in vast areas is less than 1 meter, these vessels are operative over the entire central polar basin. Moreover, the Typhoon's missile deck is elevated; by breaking through the ice this feature keeps the missile ports from being blocked, as the ice slides off (see Figure 1). A more level construction, as on the American Polaris submarines, will often have its ports blocked by ice. The elevated missile deck on the Delta vessels, formerly thought to be so as to make room for missiles, is understood today to have partly the same function as on the Typhoons.[3] Thus some 60 percent of the strategic submarines of the Northern Fleet have unique features particularly enabling them to perform sub-ice operations. (see Figure 1).

Figure 1. The Elevated Missile Deck of the Delta-Class Submarines.

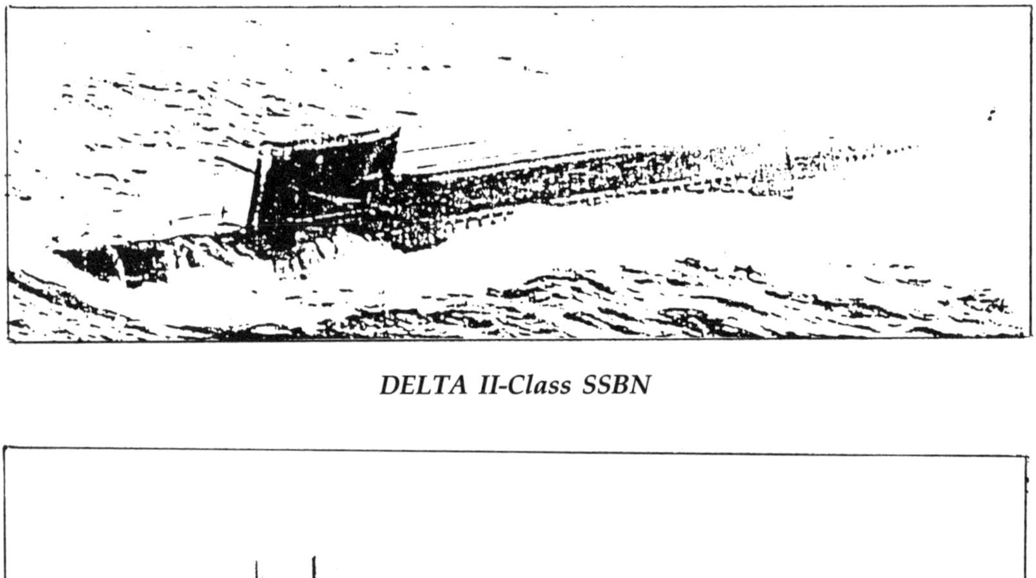

DELTA II-Class SSBN

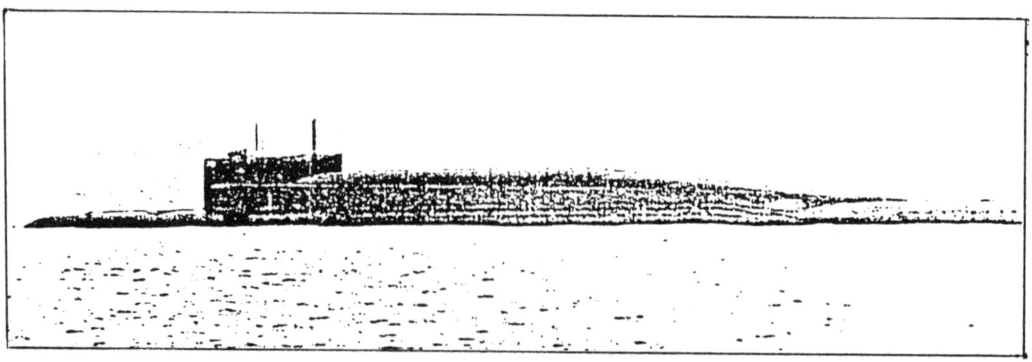

DELTA III-Class SSBN

Source: Soviet Military Power, Department of Defense, Washington, D.C., May 1981.

This change in Soviet deployment strategy was indicated by Admiral Hyman Rickover to Congress in 1975; in his opinion, the Arctic Ocean was a future strategic operational area for submarines.[4] Today the American Navy appears quite confident of this. Admiral James D. Watkins, U.S. Chief of Naval Operations, recently stated that "the Soviets are demonstrating a strong interest in operating under the ice. We better be able to know how to fight them in that region."[5] This is the reason for Congress' current consideration of a new submarine designed for hunting and fighting Soviet strategic submarines under the Arctic pack.[6] The Pentagon now seems to acknowledge what the Soviets have done for a long time: the Arctic is situated in the center of the industrialized world rather than in its periphery. The great industrial centers of North America, Europe, Japan and the Soviet Union are all within 7,000 km of the North Pole—more than 80 percent of the world's industrial production is taking place to the north of 30 degrees and some 70 percent of the world's million-metropoles are within the same area. The Delta and the Typhoon submarines can reach these from a firing position in Arctic waters (See Table 1).

These factors point toward a considerable militarization of the sub-ice waters in the next decades by both superpowers. This is a new turn in international politics and strategy that exerts a potentially large influence on the general course of events in the area. The submarines, however, are not alone in seeking refuge in the Arctic; oil rigs are already in place.

Table 1. The Distance From Two Positions in the Arctic Ocean to the Center of Selected Cities in the USA (Error Margin 1km).

Launch position	Target	Distance in km. from the launch position to the center of the target
South Cape	New York	5,650
(A small island south of	Key West	7,570
Spitzbergen)	Los Angeles	7,350
Position: 16° 32′53″E,	Seattle	5,900
76° 28′47″ N	Chicago	5,880
North of Banks Island	New York	4,450
Position: 120°W, 75°N	Key West	5,990
	Los Angeles	4,560
	Seattle	3,020
	Chicago	4,030

The Search for Hydrocarbons

The search for petroleum goes on along the entire Arctic periphery—in Alaska and Canada from the Mackenzie delta over the islands to Labrador, on Greenland and Spitsbergen and further east to the Bering Strait. In Alaska production has already commenced; Prudhoe Bay deposits may contain as much as one-third of the known reserves of the United States. The first concessions on the shelf were allocated in December 1979, and test drillings in the area have yielded highly promising results.

In the Canadian Arctic the search has been concentrated to the Mackenzie delta and the northern islands. Over the last 20 years the Canadian oil company, Panarctic Oils Ltd., has invested more than $400 million in oil exploration on the islands. Official calculations estimate some 35 percent of the potential gas reserves of Canada to be located there.

The deposits in the Soviet Arctic are of gigantic proportions. In the Tjumen district 130 deposits of oil and gas have been discovered, and several of these are probably among the largest in the world. Soviet experts have calculated production in this area to cover some 50 percent of the Soviets' total oil consumption. Also, in the continental shelf to the northeast of the Kola peninsula in the Barents Sea, wells have been drilled. The petro-geological structures in the Barents, Kara and East Siberian Seas have proved favorable. In 1975 gas deposits constituting one-eighth of the Soviets' gas reserves were found on the Yamal Peninsula; moreover, geologists have indicated that the shelfs in the Laptev, the East Siberian and the Chukchi Sea may contain 2-5 times as much petroleum as the mainland.[7] According to Soviet sources, by January 1, 1984, there were more than fifty petroleum structures in Soviet Arctic offshore areas. Of these, 17 were located in the Barents Sea and 2 in the Kara Sea, totalling 7,355 sq. km. that have been prepared for exploratory drillings.[8]

These developments are straining the limits of economic, technological and human capacity—the Arctic is the ultimate borderland for human activity. What then are the driving forces behind these endeavors?

Driving Forces in the Conquest of the Arctic

In a short paper it is not possible to give an exhaustive answer to this question. We shall, therefore, restrict our discussion to the relationship between specific actions and relations in international politics in general, viewing these as some of the more important explanatory factors.

Military Action-Reaction

According to western analysis, the Northern Fleet has several tasks. It is usual to distinguish among five different assignments:

1. contribute to the maintenance of the terror-balance
2. neutralize western aircraft carriers
3. cut off the support lines between the United States and Europe
4. back up landing operations and the army
5. boost Soviet foreign policy in general[9]

The first and last of these tasks are attended to in war as well as in peace, while the others apply to wartime situations only. To fulfill its function, the Northern Fleet had to control the Norwegian and Barents Seas to allow their submarines passage through the GIUK-gap (see Figure 2).

When the Americans invested vast amounts of capital in new weapons and operational systems[10] (in order to be able to perform a barrier strategy in this area), a potential Soviet advance was intercepted; the Yankee-class submarines risked being put out of operation before reaching their firing positions to the American east coast. In order

Figure 2. The North Polar Region and The GIUK-Gap.

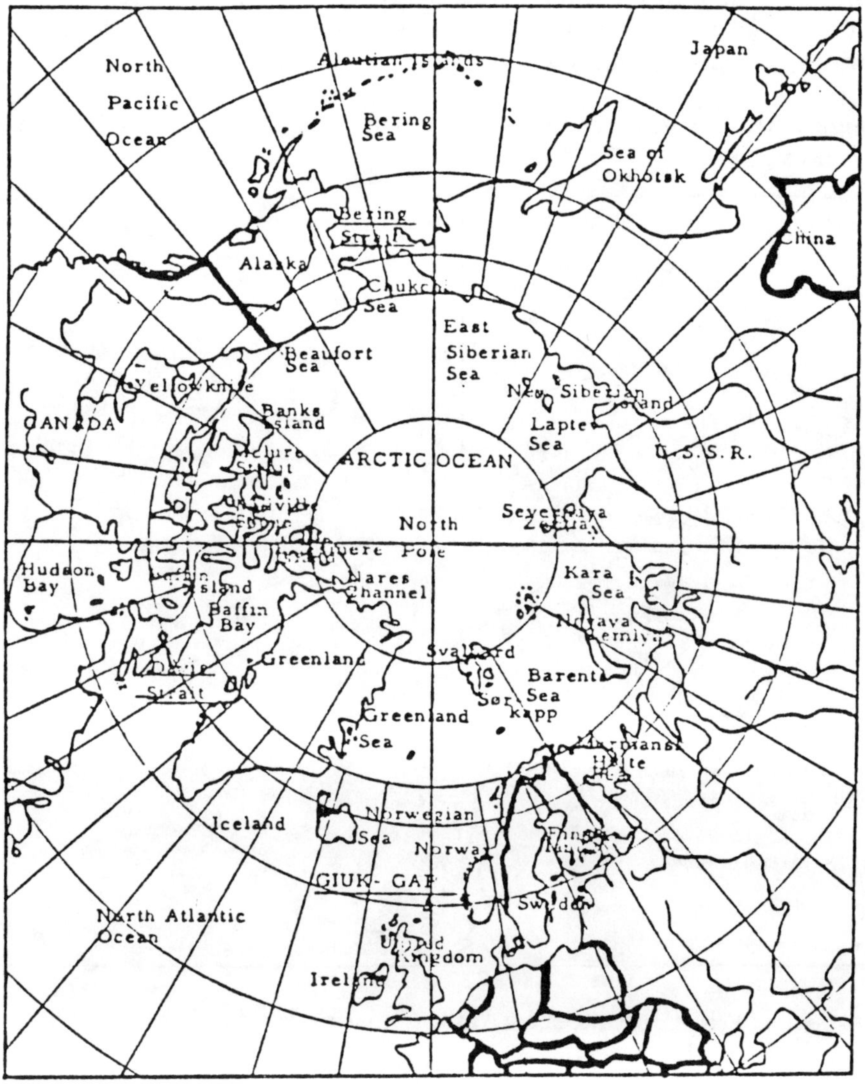

to retain their retaliatory capability, the Soviets had to counter this threat. This was accomplished by deploying first the Delta and later the Typhoon submarines, thereby avoiding having to sail into hazardous waters.

The Soviets are well aware of the problem their submarines face in this area; they have extensive knowledge of the acoustic sensing systems in operation and of the new ones being developed.[11] The detailed information the Soviets hold was revealed in an article in the Moskov Sbornik in 1976, where the author asserts the United States will cover the Denmark Strait with Captor mines and that nine SOSUS installations, with hundreds of hydrophones on each, exist in the GIUK-gap.[12] The author's conception of western strategy in this area is depicted in Figure 3A. The Soviets' unfading interest in this issue is also demonstrated in a book published in Moscow in 1980, in which the central assumption is that the American Navy will attempt to deny Soviet submarines passage through the GIUK-gap by mining the area.[13]

Figure 3A. Soviet Perceptions of the Barrier-Strategy in the GIUK-Gap.

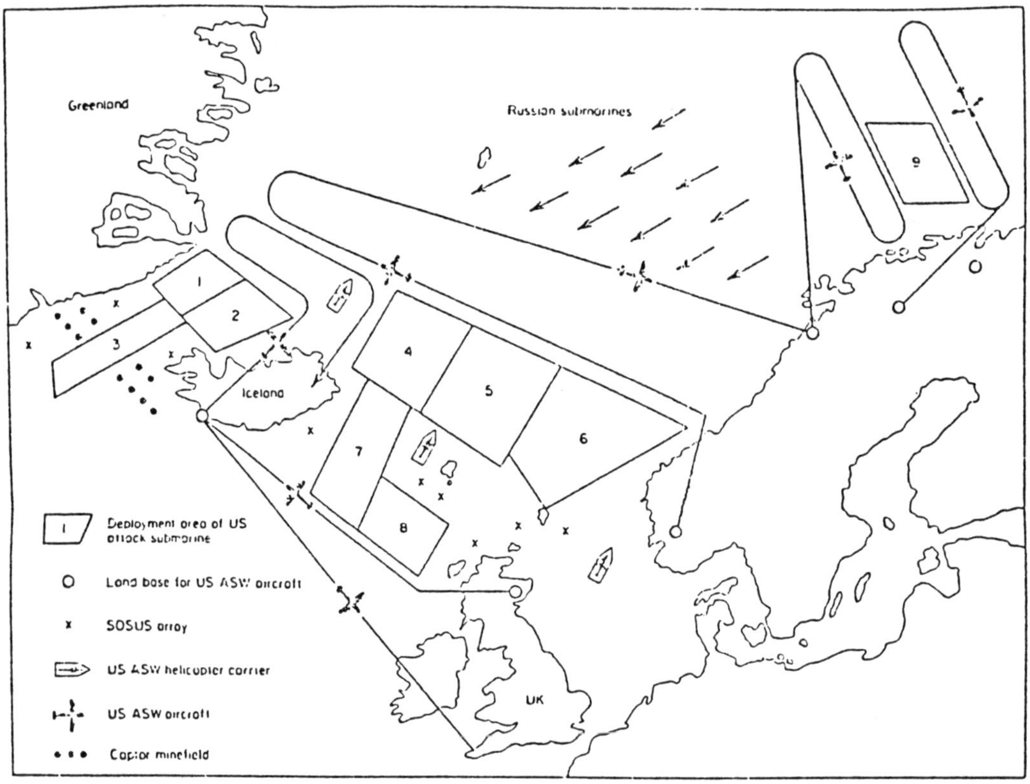

Source: Joel S. Wit, "Advances in Anti-Submarine Warfare" in *Scientific American*, February 1981, p. 41.

The withdrawal of the Soviet Union's strategic submarines to Arctic waters was a direct consequence of the US/NATO barrier strategy in the GIUK-gap. Soviet admirals are compensating for a military problem in the south by countermoves in the north. The question then, is whether this is a permanent change in Soviet strategy, having an enduring influence on developments in the Arctic Ocean into the next century.

The Americans have already given notice that they will follow suit and try to neutralize Soviet submarines in the Arctic Ocean. We are thus facing a new potential action-reaction pattern which might induce the Soviets to consider the option of leaving this area again. We do not hold a Soviet reaction of this character to be likely, however, as the weapons technology and the oceanographic conditions (pertaining to the ice layer) point in another direction. In a nutshell, the situation is that it is easier to hide a submarine than to find it under the ice. Tracking down vessels by acoustic devices on the seabed is rather difficult as the submarine-generated sound is drowned by the noise produced by cracking ice. Computers analyzing this noise often have difficulty assessing whether there is a submarine present or not. Similarly, satellites that sense by infra-red radiation are not effective, as this type of radiation does not penetrate ice. Polar darkness in winter and frequent fog in summer complicate visual detection from aircraft. In addition, there are limits posed by the environment; surface vessels designed to track submarines are not constructed for ice-breaking and have to restrict their operations to areas bordering the ice.[14] The only effective system for tracing and destruction is the so-called attack submarines, the purpose of which is to render the strategic submarines harmless. Every submarine search is a coordinated effort in which several weapons systems collaborate. Under the polar ice, however, attack submarines are left to themselves in the chase. In this way the ice canopy constitutes a great strategic asset to the Soviet Union; to the United States, it is a problem of corresponding magnitude.

Figure 3B. Submarine Operations Beneath and Through the Ice.

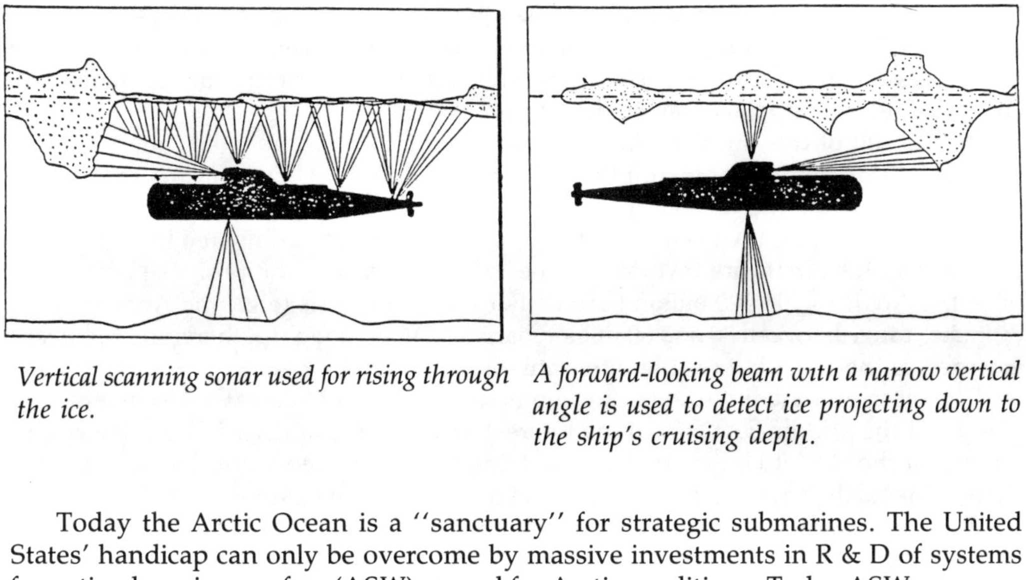

Vertical scanning sonar used for rising through the ice.

A forward-looking beam with a narrow vertical angle is used to detect ice projecting down to the ship's cruising depth.

Today the Arctic Ocean is a "sanctuary" for strategic submarines. The United States' handicap can only be overcome by massive investments in R & D of systems for anti-submarine warfare (ASW) geared for Arctic conditions. Today ASW poses extreme problems even in ice-free waters, and it is probable that development of suitable ASW systems for Arctic conditions will require several decades. Meanwhile, the Soviet Union is improving its operational ability and technology. Hence, the developments we are only now beginning to understand will probably be amplified and stand out on the strategic scene in the Arctic Ocean in the next century.[15]

Economic Action-Reaction

The world experienced its first real oil crisis in 1974 and the international oil market was severely upset. Until then the modern world had been accustomed to an abundance of cheap oil—the most important and, in some cases, the only source of energy. This one-sided dependence rendered the industrialized countries vulnerable, and the OPEC countries took advantage of this weakness to influence the outcome of the October War. On October 16, 1973, OPEC for the first time unilaterally fixed the price on oil, which soared to an average $5.12 per barrel, a 100 percent increase. The Arab member states also agreed to an immediate 5 percent reduction in the production of oil and further reductions in the ensuing months until an acceptable outcome of the war emerged.[16] At the time, the consequences of such drastic measures were not fully realized by the industrialized world:

- OPEC had become a strong, operative organization with enormous political and economic power
- the Arab members of OPEC were able to agree to and carry out an oil blockade, partly by production reductions and partly by embargo against some countries
- OPEC could unilaterally fix prices without considering the opinion of oil companies. This was a new trend; prices now increased by larger leaps.

The abundance of oil suddenly was a historical phenomenon, the international oil system underwent a change of dramatic proportions, having a permanent effect on the organization of the oil market. The change was cemented by ensuing events, the fall of the Shah in 1978 and the various crises in the Persian Gulf. These events caused large price hikes, amounting to 1,000% in the period 1973–79 which, in turn, contributed to the recession of the 1970s and 1980s. The vulnerability of western countries to disturbances in their oil supplies was evident to all.

As a result of this, industrialized countries poor in oil accelerated their efforts to reduce their dependence on supplies from the Middle East. One means to this end among several was the development of new production areas, mainly in the North Sea, Mexico, Southeast Asia and Africa, but also in the Arctic. Compared to other areas, prospects in the Arctic are favorable; sober estimates assess the probable amount of oil in the Arctic to 300-500 billion barrels. For the other areas (e.g., the Antarctic) the estimates range between 30 and 60 million barrels.[17] In the long run therefore, the Arctic emerges as the most interesting alternative to the Middle East.

The oil crises had different ramifications for the Soviet Union. The decline in supplies from the Middle East stirred the interest of western countries in petroleum supplies from the Soviet Union, thereby enabling it to increase its international market share. This might have induced the Soviets to increase their exploratory efforts in the Arctic. In particular there was and is an acute need for new wells; growth in the country's oil production has recently been some 7 percent annually, while domestic consumption has on average grown some more. Increasing consumption in Eastern Europe and other countries supplied with Soviet petroleum adds to this need for an expansion of oil production. By the end of this decade the present pattern of consumption can only be upheld by imports or by supplies from new deposits.[18] It is ample reason to believe that the Soviets will accelerate their efforts to get new wells into production. In addition to the Arctic mainland, the Barents and Kara Seas emerge in the short run as the most likely production sites.

Thus, East as well as West have foreign-economic and foreign-political reasons to expand oil and gas production in the years to come; the West in order to reduce its vulnerability to market instability, the East to maintain the economic and political advantages it reaps from oil production.

Action-Reaction: From the Specific to the General

Both the economic and military issues discussed display the following features:

- The exploitation of the Arctic is a direct consequence of and is strongly influenced by problems in southern latitudes
- The exploitation of the Arctic is intended to ensure vital national interests—it has to do with high politics

The first feature is significant because it demonstrates that the exploitation of the Arctic is not motivated solely by narrow economic or military-industrial interests. Rather, it is an integrated part of international politics and represents a response to problems that are not immediately solvable. There is nothing in American naval strategy that indicates the United States is about to leave the North Atlantic. On the contrary, the new American naval strategy of forward deployment implies that the military effort will be stepped up, not only in the GIUK-gap, but also in the northern areas outside the bases on the Kola peninsula.[19]

The economic recovery we are now witnessing also points to increased activity in the Arctic, as demand for petroleum probably will increase due to the economic upswing. According to IMF sources there are five major influences at work:

> ". . . first, the efficiency achieved in recent years in the use of energy has reached the stage of diminishing returns. Second, future conservation measures may be more difficult to realize because consumer countries have saved energy the last years at the expense of capital and other labor resources. Third, there are growing uncertainties regarding new sources of supplies. Fourth, the economic and political costs of alternative sources of energy are rising. Fifth, the oil market is potentially more volatile. The spot market, a residual market accounting for less then 5 per cent of the transactions in the 1970s, may now count for as much as 40 per cent of oil trade in the non-communist world"[20]

All this leads a number of energy analysts to believe that there may be another explosion of oil prices in this decade. Many past scenarios of the experts have proved false, but a wise person takes out an insurance policy. These factors support the view that present developments in the Arctic will continue toward the turn of the century.

The second feature, pertaining to high politics interests, indicates the strength of this development. Due to the high politics aspect none of the implied states can afford to relax its attendance to these interests. The relationship between stable and secure petroleum supplies and national security is an obvious one, as stated by the American Department of Finance:

> ". . . any sudden disruption (in oil imports), in excess of one million barrels a day, would have a prompt and substantial effect upon our economic well-being, and, considering the close relation between this nation's economic welfare and our national security, would clearly threaten to impair our national security."[21]

In the same report the Department of Defense emphasizes that:

> "The degree of our energy self-sufficiency must be such that any potential supply denial will be sustainable for an extended period without degradation of military

readiness or operations, and without significant impact on industrial output or the welfare of the populace.''[22]

Corresponding points of view are held also by the other Arctic states.[23] As a consequence of the price increases in the seventies, and those likely to come, the oil companies should have the necessary incentives to intensify their exploration and exploitation of petroleum resources in the cost- and technology-intensive Arctic in the decades ahead. The security/political concerns of governments and the economic motivations of oil companies point in the same direction: northwards.

A powerful alliance has emerged and, as with most powerful constellations, this merger between governmental and industrial interests is not necessarily sensitive to the concerns of others. This point is underscored by the fact that this alliance has been formed simultaneously in several countries, all attempting to ensure their own national needs and interests in competition with each other. This multinational satisfaction of demands occurs in an area where some vital questions of sovereignty are unresolved.

The Law of the Sea in the Arctic: Who's in Charge?

Three issues are relevant in this connection: the lack of sea-borders, sector claims, and the absence of an international law of sea-ice.

Geographically, the Arctic Ocean is a boundary sea between East and West, and between five individual states. There is, however, only one boundary that is conclusively drawn, viz. between Canada and Greenland in 1973. The following boundary issues still await resolution: the delimitation line between Norway and the Soviet Union in the Barents Sea, between the United States and the Soviet Union in the Chukchi and Bering Seas and between Canada and the United States in the Beaufort Sea. These issues seem difficult to resolve as they touch upon the military-economic interests referred to, and because of the sector principle invoked by some parties.

Two countries are maintaining the sector principle as basis for their claims to sovereignty over lands and islands in the Arctic Ocean: In 1925, Canada claimed sovereignty to all land in the Arctic Ocean between the mainland and the North Pole in the sector between 61° and 141° west. A year later the Soviet Union claimed sovereignty over all land—discovered and undiscovered—within the sector between 30°04′35″ east and 168°49′30″ west in the Arctic Ocean. These claims pertained to land and islands only, and not to the column, the ice, the ocean floor or the airspace above the sectors (see Figure 4). However, it is a matter of speculation whether the claimants intend to extend their claims to encompass these other elements.

In Canada this issue was discussed in public in 1956, when a Soviet ice station drifted into the Canadian sector without receiving a protest from Canadian authorities. The parliamentary opposition alleged the Soviet presence to be a violation of Canadian sovereignty and that the government ought to act accordingly. The government assessed that the Canadian claim applied to the islands only and not to the sea—a protest was therefore out of the question. It was, however, emphasized that the potential use of the ice justified a Canadian exercise of limited jurisdiction in the area. The content of this reservation was not and is not yet specified. The only substantial demonstration of Canadian authorities' view was the Arctic Waters Pollution Prevention Act of 1970, promulgated in response to the tanker *Manhattan* sailing through the Northwest Passage. This law went further in the protection of the vulnerable Arctic environment than existing environmental protection legislation in southern areas would have provided for, and confirmed the Canadian position of having special rights and responsibilities in the area.[24]

Figure 4. The Law of the Sea of the Arctic and the Sector Lines.

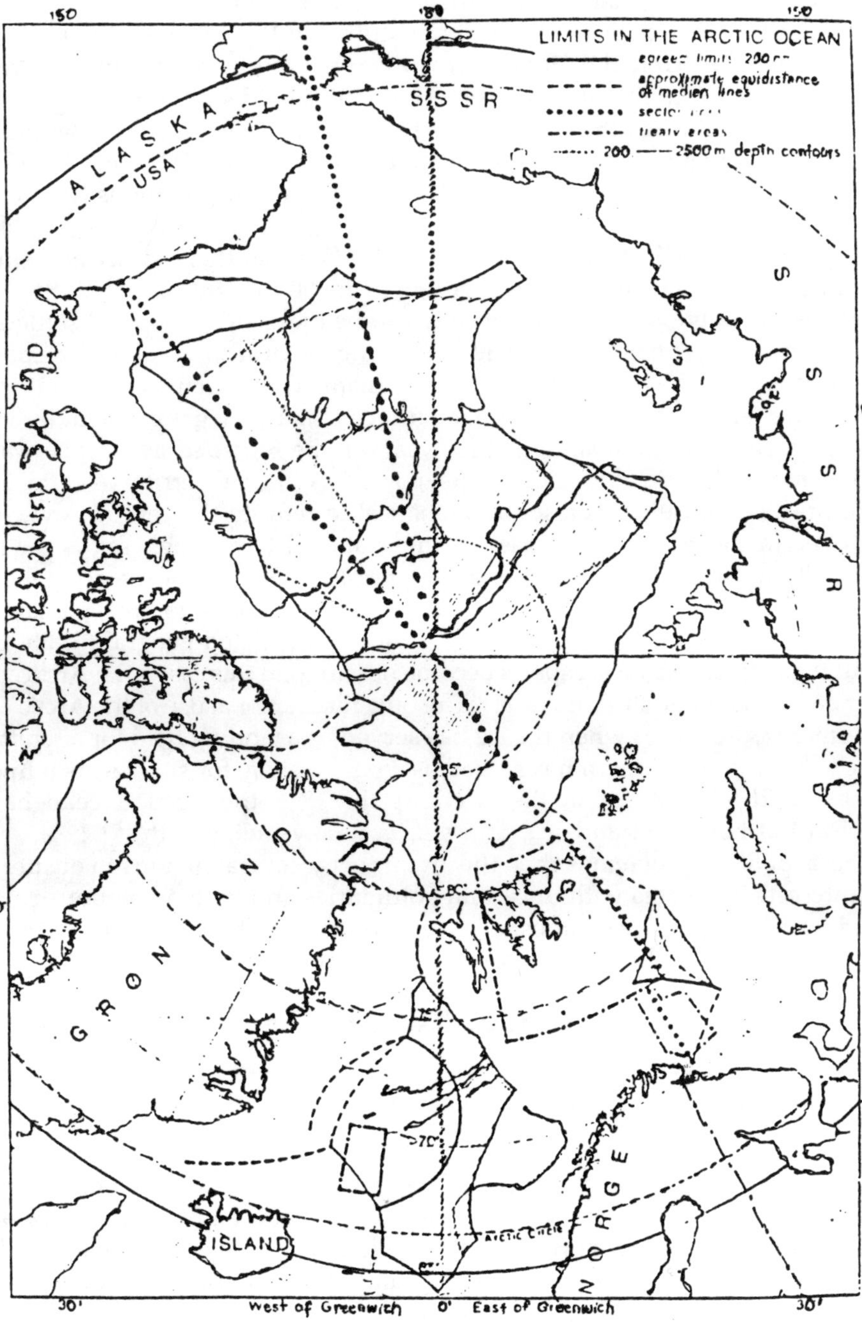

The Soviet Union has never officially voiced its intention to expand its sector claims, but Soviet legal experts have done so without being censored. Soviet authorities have the sector lines drawn as boundaries for "Soviet Arctic Areas" on their maps. From this and other evidence it has been surmised that Moscow consciously keeps the door open for the possibility of extending the sector claim.[25]

In the delimitation negotiations with their neighbors, Canada and the Soviet Union maintain that the sector claims are to be constitutive in the conclusive stipulation of the boundary line. The United States and Norway reject this by repudiating the validity of the sector principle in international law. The parties hold strongly diverging views and imminent resolution of the issue seems remote. With regard to the Barents Sea, Norway and the Soviet Union have been negotiating since 1974.[26] In the Chukchi and Bering Seas the dispute is of such a delicate character that negotiations have not yet been initiated.[27]

These problems are further aggravated by the fact that there exists no generally accepted international law for the ice covering some 90 percent of the Arctic Ocean. This Ocean is the only one that theoretically can be crossed by foot and settled on a semipermanent basis. The Eskimos use it for hunting and transport, and in the postwar time the superpowers have established several manned research stations on drifting ice islands. Human presence on the ocean surface creates a legal problem: are these "islands," which often are maintained for years, to be regarded as "Arctic waters" (which they are in a physical/technical sense) or as "Arctic sea territory" (which they are in the practical sense)? It is also problematical to refer to the central Arctic Basin as a high sea in the traditional sense. In international law a high sea is delimited geographically to the ocean areas outside the territorial waters of the coastal states. Politically and juridically this refers to the rights of nations to free passage and unhindered presence. Such a high sea can easily be delimited geographically in the Arctic, but the political/juridical aspects become obscured in this case—the Arctic Ocean has never been crossed by surface vessels. The only international use of the Arctic Ocean are the numerous instances when the ice has served as a landing field for aircraft, and when the sub-ice ocean column has served as a passageway for submarines. In other words, the juridical problem in the Arctic is that "...the Arctic Ocean has the characteristics of both land and sea, and yet, is totally unlike both."[28]

The basic political problem is thus the attempts by several nations to ensure their political interests in an area with uncertain boundaries and with ambiguous rules for state conduct. This explosive composite leaves the states with two options: either seek a compromise in an area where interests are colliding, or give priority to national interests irrespective of the costs that this course of action will entail. In Weberian terms then, we can construct two "ideal" future scenarios: a chaos scenario and an order scenario.

THE FUTURE: TWO SCENARIOS

Chaos

Other things being equal, no state leaders are inclined to make compromises on the security of the nation. The interests they work to ensure in the Arctic are of this kind—other political concerns normally have to yield to security considerations. International politics is the domain of the strongest, and power is essential in defending the national interest—also in the Arctic. Two types of conflict might arise: between East and West, and within the western bloc. Odd alliances might be formed; in the U.S.—Canadian conflict in 1969-70 over the right to use the Northwest Passage, Canada received support from the Soviet Union. Norway, on the other hand, has not been supported by other western countries in the dispute with the Soviet Union regarding the legal status of the shelf around Svalbard Archipelago. In delimiting the ocean area between Canada and Greenland in 1973, disagreement arose over which of the two was entitled to sovereignty over Hans' Island in the Nares Strait. These conflicts have been of a limited extent and of a controllable nature—like ripples in the water but having the potential of a veritable storm. Hence, there exists a definite need for resolving the remaining issues in a not too distant future; chaos might prevent continued economic activity in large parts of the Arctic Ocean.

Order

When a new territory like the Arctic Ocean becomes the target of the industrialized world's effort to satisfy its needs, it is hard to conceive of a development without conflict. The essential point is to maintain the conflicts at controllable levels. In the chaos scenario, all other concerns yield to national high politics interests. Acting from this premise, a state must necessarily harm others if its respective interests are incompatible, thereby creating international conflicts. The purpose of security politics, however, is the prevention of conflicts and the avoidance of war. Hence, taking the interests of other states into consideration is an important element in every sensible country's security policy; one cannot afford actions that jeopardize one's own interests. This can be illustrated by the following issues: the possibility of Canada and the Soviet Union extending their sector claims to comprise all elements within their unilaterally-proclaimed sectors.

Strategically, the option of extending sector claims is dubious; for reasons of maneuverability, strategic submarines are forced to operate in the central polar basin that, in places, is as much as 5400 meters deep. The Soviet shelf in the Arctic is extremely shallow; 75 percent of it is less than 100 meters deep.[29] That means that most of this shelf, which covers some 5 million sq. km., is too shallow to provide operational space for submarines between the ocean floor and the ice. To the Soviet Navy then, the optimal status of the area north of the continental shelf margin is that of high seas (in the conventional sense of the concept) since the Soviet sector only encompasses a slice of the central polar basin. From a military point of view, it is not a desirable option to limit the operational space to this slice, as American attack submarines then get highly favorable conditions for performing their anti-submarine warfare (ASW) missions. Regarding the central Arctic Ocean as a high sea has been in accordance with the Soviet position on ocean law-related issues since the 1970s. Witness the Soviet establishment of a fishery zone in the Arctic Ocean in 1977 and the economic zone established this year.[30] In both cases, they have operated within the framework of the existing ocean law as it has evolved during the UNCLOS III. The optimal solution for the Soviet Union regarding the Arctic Ocean—in terms of its military and economic interests—is to have sovereignty over the entire shelf to the abyssal plain, a 200-mile economic zone and unrestricted operational conditions ensured in the central Arctic basin. Hence, Soviet policy in this area is within the confines of international law.

The Canadian Arctic Water Pollution Prevention Act of 1970, is supported by the new Law of the Sea Convention of 1982, article 234, which states:

> "Coastal states have the right to adopt and enforce non-discriminatory laws and regulations for the prevention, reduction and control of marine pollution from vessels in ice-covered areas within the limits of the exclusive economic zone, where particularly severe climatic conditions and the presence of ice covering such areas for most of the year create obstructions or exceptional hazards to navigation, and pollution of the marine environment could cause major harm or irreversible disturbance of the ecological balance. Such laws and regulations shall have due regard to navigation and the protection and preservation of the marine environment based on the best available scientific evidence."

Thus Canada and the Soviet Union, both having signalled an alternative view on the legal status of the Arctic Ocean, seem to be in a situation where their interests are optimally attended to within the framework of the new Convention on the Law of the Sea. However, all problems are not resolved, it still remains to work out an international law of the ice. Most of the boundaries between adjacent states await their stipulation, and economic and military interests are still to some extent incompatible.

In order to transcend this deadlocked situation and thereby render the various problem-complexes more solvable, a division of the Arctic Ocean into zones according to types of activity and sovereignty relationships might be instrumental. A zonal arrangement serves to isolate the various types of activity from each other, thereby simplifying the sovereignty issues and maintaining conflicts at a controllable level.[31]

A Zonal Arrangement for the Arctic Sea: Activities and Jurisdictions

The need to separate the various types of activities can be met by dividing the Arctic Ocean into three zones: a national Arctic, an intermediate Arctic and a high Arctic (Figure 5). The national Arctic geographically encompasses all land, islands and internal waters north to the polar circumference, and is subject to unrestricted national jurisdiction according to international and national law. Hence, it is the business of the respective national governments to decide on the type of activities that can be performed in this zone. Today the activities here are rarely internationally controversial; it is in the other zones that jurisdictional functions and various activities may collide.

The intermediate Arctic includes the areas where restricted but recognized national rights are in force, viz., the shelf areas, the economic zones and the territorial waters. The intermediate Arctic thus covers the area between the base lines and the shelf margins' outer limits of the economic zones. The limited jurisdictional powers awarded the national states in these waters implies that other countries might operate here if the rules laid down by the holder of sovereignty are complied with, according to Article 234 of the New Convention on the Law of the Sea. Since this zone can be used by several states, both economic and military activities must to some extent be allowed here. It is, however, reasonable to assume that the zone will harbor mainly economic activities, as the military ones will largely be performed on land or in the central polar basin. The contemporary military use of this zone is that of annual convoys along the northern sea route between Vladivostok and Murmansk. Submarines operating in the White Sea and in the central polar basin will have to transit the Greenland and Barents

Figure 5. The Proposed Arctic Zones.

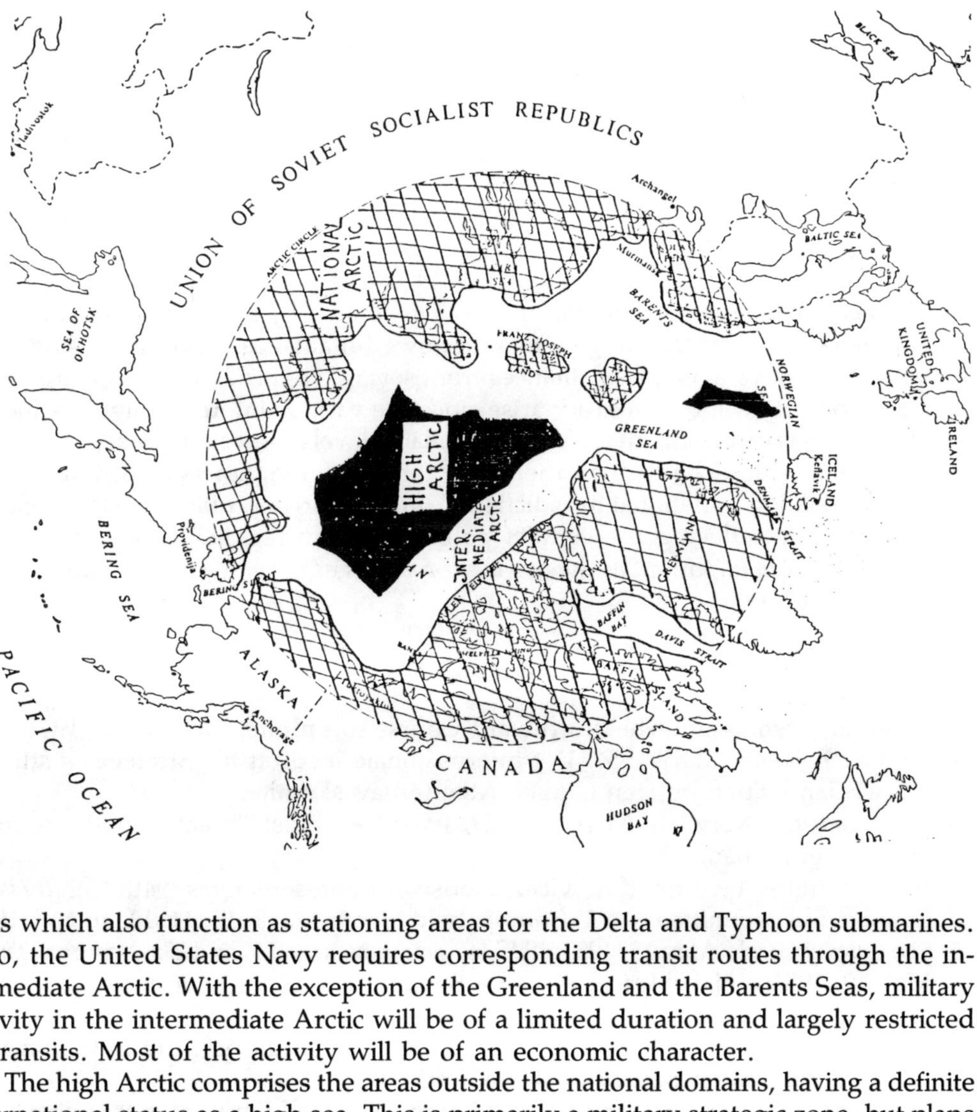

Seas which also function as stationing areas for the Delta and Typhoon submarines. Also, the United States Navy requires corresponding transit routes through the intermediate Arctic. With the exception of the Greenland and the Barents Seas, military activity in the intermediate Arctic will be of a limited duration and largely restricted to transits. Most of the activity will be of an economic character.

The high Arctic comprises the areas outside the national domains, having a definite international status as a high sea. This is primarily a military strategic zone, but plans exist for constructing sub- and surface supertankers intended for operation in the central polar basin and carrying petroleum from production sites in the Arctic to markets in the United States, freight routes—the sub-ice ones in particular—should be channelled into certain regions so as to avoid collision with military interests.

This arrangement is to be regarded as both *instrument* for initiating a process to resolve the remaining legal issues *and* as a permanent, orderly system regulating various types of activities in the Arctic.

Figure 6. Types of Activity in Different Zones.

		ZONE		
		National	Intermediate	High
ACTIVITY	Economic	X	X	
	Military	X		X

SUMMARY

The need to ensure high politics interests generated by developments in international politics has given rise in general to a powerful alliance: government-industry-military. When several of these alliances from several countries work to realize their interests, conflicts will by necessity arise since the various interests might be incompatible. Conflicts can be maintained at controllable levels, however, by separating the various activities and channelling them into three different zones according to their nature and relation to national jurisdiction. Thus, sensitive problem areas are clearly defined, thereby reducing the conflict potential in the region. Regulating activity becomes ever more important as the driving forces discussed gain strength, accelerating the course of events.

NOTES

1. Viljamur, Stefansson: The Northward Course to Empire, New York, 1922.
2. Militaer Balansen 1983–1984, Det internasjonale institutt for Strategiske studier, Norwegian edition by Den Norske Atlanterhaveskomité.
3. *Conn. Bulletin,* Norwich, November 29, 1981: "A Soviet Threat from the Arctic," by William Parham.
4. See Committee on Armed Services, House or Representatives, 94th Congress, 1. Session, Subcommittee No. 2. *Hearings* held March 4, 6, 8, 10, 11, 13, and 17, 1975.
5. See *Calgary Herald,* May 15, 1983: "US ice-war subs urged," and *The New York Times,* May 19, 1984: "Navy Trains to Battle Soviet Submarines in Arctic."
6. *Ibid.*
7. See Willy Ostreng: *Polhavet i internasjonal politikk,* Study AA: 012 in the Nansen Institute Publication Series, 1978 chapter 5.
8. Tsjingiz G. Sattarov: *Statement on the Current State and Prospects of the Oil and Gas Development on the Outer Continental Shelf of the USSR,* Talk at the Nansen Institute June 14, 1984, p.4.
9. See John Kirsten Skogan and Anders Sjaastad: *Politikk og sikkerhet i Norskehavsomradet. Om de enkelte land og vare felles problemer, Oslo 1975.*
10. See Willy Ostreng: *The Soviet Union in Arctic Waters. Security Implications for the Northern Flank of NATO,* Occational Paper, Law of the Sea Institute, Honolulu, 1984, chapter 3.

11. See B. Tepelinsky: ''America's Naval Programmes,'' in *Survival,* March-April, 1973, pp. 83–84.
12. Jim Bassert: ''Computers and New Effectiveness to SOSUS CAESAR in *Defence Electronics,* October p. 63.
13. *Omako World Herald:* ''To Bottle up Submarine Fleet. Russia Expects Mine Curtain,'' October 14, 1980.
14. See Supranote 10, Chapters 4 and 7.
15. *Ibid.,* Chapter 7.
16. See Jens Evensen: ''Oljekrisen og norsk oljepolitikk,'' talk in Oslo Arbeidersamfunn 4.12.-79.
17. See Anders Kjolberg: ''Oljen i Arktis. Muligheter, problemer og utfordringer, in *Internasjonal Politikk,* no.4, 1977.
18. J.P. Stern: ''Western Forecasts of Soviet and East European Energy over the Next Decades (1980–2000), in *IEC* (1981).
19. John Lehman: ''War against Subs: The West Hones its Skills,'' by Drew Middleton. *The New York Times,* January 16, 1983.
20. See Jonathan Power: ''A New Oil Roller Coaster? Apply the Brakes Now,'' in *International Herald Tribune,* September 13, 1984.
21. David R. MacDonald: US Department of Treasury Memorandum, ''Report of Investigation of the Effects of Petroleum Imports and Petroleum Products on the National Security Pursuant to Sec. 232 of the Trade Expansion Act, as Amended'' 13 January 1975.
22. *Ibid.*
23. See for instance Independent Research Associates: ''The Strategic Importance of Arctic Oil,'' article in The Canadian Institute of International Affairs' bibliotek, Toronto 1970. See also Erik Hesselbjerg: Greenland. ''Oil—But Not At Any Price,'' i *Danish Journal,* 76/1973, and T.A. Hockin & P.A. Brennan: ''Canada's Arctic and Its Strategic Importance,'' i E.J. Dosman: *The Arctic in Question,* Toronto, 1976, *Dansk Energipolitikk 1976,* Handelsministeriet, Kobenhavn, 1976.
24. See Bo Johnson Theutenberg: *The Evolution of the Law of the Sea: A Study of Resources and Strategy with Special Regard to the Polar Areas,* Dublin, 1984, pp. 26–61.
25. See S.M. Olenicoff: ''Territorial Waters in the Arctic: The Soviet Position,'' R-907 ARPA, July 1972, The Rand Corporation.
26. See Willy Ostreng: ''Norwegen und die Sowjetunion in der Barentssee,'' in *Europa-Archiv,* 25. Jahr. vol. 23, 1980.
27. Robert Krueger: ''Bering Sea Petroleum: A New Meeting Ground by the Soviet Union and the United States,'' paper presented at the Second Workshop of the Geology and Hydrocarbon Potential of the South China Sea and Possibilities of Joint Development, August 22–26, 1983, East-West Center, Honolulu, Hawaii.
28. See supranote 25.
29. See supranote 8.
30. See Willy Ostreng: ''Delimitation Arrangements in Arctic Seas: Cases of Precedents or Security of Strategic and Economic Interests?, Paper presented at the 18th Annual Conference of the Law of the Sea Institute in San Francisco, September 24–27, 1984. To be published in *Ocean Development and International Law,* winter 1984/85, and in the Conference Proceedings: *The Developing Order of the Oceans,* Law of the Sea

Institute, Honolulu, Hawaii, 1985/86.

31. A zonal arrangement of a partly different kind has previously been proposed. Finn Sollie: ''The Need for an Arctic Basin Treaty,'' unpublished paper in the Nansen Institute. See also same author: ''Forurensing av polaromradene? Et sporsmal om regulering av utviklingen'' in *Internasjonal Politikk,* no. 1, 1972.

The Broad Perspective

A.E. Pallister
Pallister Resource Management, Ltd.

This is a tough assignment, to give a broad perspective of the Arctic. We often talk about the Arctic as if it were finite. As we attend conferences we learn that it has very many variabilities, being such a large area. It is like undertaking a broad perspective of South America or North America. I will attempt to look at the Arctic with the combined technical, social, economic, and political aspects in mind, but I would like to focus on the restricted subject that has been prominent in our discussions in the last three days—offshore oil and gas. I am able to do that with ease, since Willie Ostreng has covered the political and international aspects so well. So, I will comment on offshore oil and gas specifically, as the "locomotive" that is going to drive Arctic engineering in the 21st century, the topic of our Symposium. In doing so, I offer my projections in four categories.

Firstly, I would like to challenge our thesis that offshore petroleum production will take place in the near future. Secondly, I will sum up what I have interpreted as being exposed at this conference in terms of where we are technologically today. How prepared are we to obtain oil production from offshore prospects north of the Arctic circle, more than a few miles from shore? Thirdly, I would like to speculate. This is very dangerous, but I think somebody has to do it at this time! In the spirit of this Symposium I will speculate as to what could happen in the future. Finally, I will come back to "policy"—the Broad Perspective—and comment on the need for collaboration among governments, industries and professional societies.

ARCTIC OFFSHORE PETROLEUM PRODUCTION

Are we going to have oil and gas production from the offshore in the Arctic? And when? We need to recognize that we come to this symposium as a biased group. Most of us have an interest or an expertise in engineering for offshore petroleum production. It is natural that we start our discussions with the assumption that this development will happen. I am sure that in the minds of our assembly there is a rationale of why it should transpire, so I pause a moment to remind you that the Arctic is a place which has been the object of a lot of ideas and concepts, but not very much has happened. It has been an area of dreams with "booms and busts." I really wonder what is so different now to make us think that in the 21st century the time will arrive for industrial development.

Let us examine the Arctic for a moment in the most simple terms. It is different from the rest of the world because it is remote from populated communities, and it has a unique climate. These two facts introduce a large number of constraints to industrial development. In the offshore, and we keep coming back to the fact, the sea is covered with ice.

Equally fundamental, the region is cold. It is also dark for long periods. There are very few people and little infrastructure. Also, there is an important element we don't talk about very much: The public attitude towards the Arctic that it should be left the way it is—a pristine Arctic, a model of conservation and environmental protection. All these basic factors add up to high cost. In terms of oil and gas, notwithstanding short periods when the industrialized countries have problems of delivery, oil has been in over-supply throughout this industry's history. We do have alternatives for Arctic offshore oil and gas. We have undiscovered and underdeveloped resources from the rest of the world to draw upon. And, the southern climes are a much easier place in which to work.

I am stressing the offshore, since technically and operationally the onshore presents no particular problem. It is no colder in Prudhoe Bay in January than it is in Minneapolis. In fact, sometimes it is warmer. But, the offshore is a different matter altogether. There will be many new requirements. We have heard a lot of them here, and I judge we are just seeing the tip of the iceberg. When we get into details of the requirements of operating fulltime offshore, and I emphasize, "operating" (not exploring, with its flexibility), new technologies and techniques will be needed. We need to do a lot of work to make sure we can operate efficiently and safely.

There is a significant difference between exploration for and production of oil and gas. Exploration is motivated by concepts and images of finding the elephant, the 8 billion barrel oil field. We are drawn along by the fact that one has been found, on land, at Prudhoe Bay. That is what is motivating the presently large expenditures in the offshore. Hope, and confidence in the future. But, when we get to the production stage more than a few miles from shore, the basis for expenditures, which are in the billions of dollars category, will be on the hard facts of economics—not on visions.

What are the new requirements for producing from the Arctic offshore? One could fill a screen with all the new requirements, but I would like to group them within two "generic" terms. One is safety, the other is economics. Unfortunately, those two terms are contradictory to some degree. Safety requires particular attention in response to the attitudinal aspect that I mentioned earlier, the public view of the unspoiled north. While industry is giving increased attention to safe operations everywhere, in Arctic waters operations must approach fail-safe status. There are no effective means of combatting an oilspill in ice-covered waters. The oil industry, while having a statistically good safety record, does have accidents and has suffered some very serious ones in the offshore in recent years. Unlike other industries, no universal organizational system has been developed in the petroleum industry for safety management in the way the aircraft industry, for instance, has succeeded in designing safe aircraft, operated by accredited people.

In mentioning the oilspill threat, I am underlining the fact that industries and, most importantly, people on the Arctic coasts are highly dependent upon resources of the sea for their livelihood. They cannot go to a supermarket in the way we do in the south. If we have pollution problems, we have alternatives. They do not.

Speaking now on the matter of economics, petroleum from the Arctic is going to have to compete with other world-wide supplies. For a long time to come it will be cheaper to find, produce and transport oil from the Southeast Asian offshore, for instance, than from the Arctic. So, why do we assume that Arctic offshore petroleum is going to be developed?

I must present the caution that there will not be a natural evolution into the oil production stage in the Arctic offshore for the simple reasons that there are many internal operational constraints as well as external opposition. We in this room must keep coming back to the fact that many not in this room are opposed to industrial development in the Arctic on social grounds. Also, we need to recall the criteria of those who must make the huge investments to bring it about. Unless mammoth fields that are inexpensive per-unit to produce are found, the return on the investment is not attractive to the private sector. And, keep in mind that traditionally, worldwide exploration for oil has gone very fast and hard in numerous areas until the need to know has been satisfied. Prospects are determined to be economically commercial or not. The blind has come down on several areas that were once active.

Even when the elephant is found, a very solid case is going to have to be made by industry to government and by government to citizens that there is a low risk of accidents, and particularly, oilspills.

PREPARATION

Accepting that Arctic offshore resources will be developed, we need to ask another question. How prepared are we? Another characteristic of the oil industry is its adaptive nature. In the 1950s when I was first involved in Arctic oil exploration, it was a great time. We had tough work, and we did it the best we could. We fought the elements but that was all. We had no human opposition. Investment was welcomed by local residents. We hadn't heard of environmentalists or socio-economists. But, by 1970 and after, industry had been challenged.

First, came the environmental movement. It emphatically brought to our attention that there was more to consider in our activities than just finding oil. This was followed by a trend towards socio-economic analysis. Now, major projects have socio-economic and environmental impact assessments which is an industry itself. Then, the focus moved to quantifying local and national benefits. Most recently, the attention has moved to industrial and human safety. Public attitude has become a very important and influential part of the planning and operations of industry.

During the past 15 years a tremendous amount of ambient data has been gathered, much more than was ever thought to be required. Environmental data has been found to be not only helpful but essential in planning projects. A large infrastructure has been built up, particularly in government agencies, in the review and assessment process. It is now time to attenuate that growth. It is time to re-allocate these energies. Now that we have the talented and knowledgeable people, we need to direct our attention towards undertaking (not analyzing), "responsible development." Responsible development incorporates the environmental and socio-economic benefits, and safety considerations in its fabric.

In the meantime, Arctic exploration has progressed, as has been shown here in these last few days. It has been illustrated that industry can meet the operational challenges. It has also been demonstrated that it is very costly—$150 million for a dry hole in the Beaufort Sea. The jury is still out on the question of whether large petroleum reserves will be found. There have been a number of "discoveries" reported, but uncertainty remains whether the reservoirs are of sufficient volume and produceability to warrant the costs of proceeding to production. This phase introduces quantum increases of cost and technological needs.

THE FUTURE

Now, I will go where angels fear to tread, and speculate on what is going to happen in the next 50 or more years. While it is total speculation, it is based on what has happened in the past in the Arctic and where we are today. Plus, I have developed it from the discussions in the workshops and on policies of Arctic countries, as they exist today. Concisely stated, the United States is essentially a free enterprise system. There is relatively little government intervention in encouraging the exploration that is going on. It is driven by the belief that there are some commercial petroleum reserves to be found. In Canada, the situation is quite different. There are large grants, as high as 80 percent, to encourage companies to explore. As a result, there have been more than 100 wells drilled in the Arctic. So far nothing commercial has been found. Government considers that it is very important to keep activity going. In Norway, the situation in the northern waters has been "go slow." Gradually, preparations have been made and exploration activity is moving northward. In relation to the size of its massive continental shelf, the USSR has been inactive.

Taking into account all factors, what can one project? Exploration will progress slowly but safely for there are factors that keep it going. There are exploration agreements, acreage evaluations and work requirements. It will continue at about the pace of today. There will be numerous discoveries, tens of finds across this vast region. Most will be found to be non-commercial and abandoned or put aside for another day. Projects which are judged to be commercial will enter into a review and regulatory system which will require a prolonged period during which experts and others will examine all the reasons why the project should not be undertaken. A few of the finds will proceed to the production stage. These will be the large, billion barrel fields. And, we'll be in the 21st century.

(In terms of engineering, my assessment after having seen all the schemes and concepts presented at this Symposium that have been used for offshore exploration drilling, is that they can be extrapolated into the early phases of the production stage. It appears that so much progress has been achieved in exploration drilling that there is a need to digest and consolidate some of the technology rather than to develop new systems. The emphasis should be on safety and cost-reduction.)

Then production will commence, transported by sub-sea and sub-bottom pipelines to the nearest shore and then southward by pipeline. In my opinion, if there is any option of transporting oil by pipeline, it will be chosen over that of carrying oil by tanker through ice infested waters. I say this on the basis of anticipating difficulty of regulatory agencies giving approval in the face of public opposition to transporting oil through Arctic seas.

After the year 2000, with the advent of petroleum production, the time will have arrived for the use of several new technologies. Basically, these will be ice-avoidance and sub-sea systems. To date, there has been an acceptance of ice. We are building large and strong equipment to resist it or, in the case of vessels, to break through it. The new approach to working on a permanent basis in the Arctic will be ice avoidance or removal. An example of avoidance is using sub-sea installations. Ice is the major challenge. The best way to accommodate it is by avoiding it.

We will see improved economics in almost the same way, but not to the degree we have seen it with computers, for example. After going through the expensive learning phase, we will find out how to operate more economically, and the marginal fields which we have put on hold will become producible. Then, it starts getting exciting at about year 2025 when we arrive at the value-added stage such as petrochemical plants and the impact of the petroleum infrastructure in the development of other resources. Finally, I come to about year 2050, when we get into the industrial spinoff stage when secondary industries start to cluster around the primary industries and cities of the Arctic, visualized by Ben Gerwick, start to develop. Sorry to delay this event so long, Ben.

That is one scenario, the evolutionary one. There is another—the ''crash'' scenario. This one is really difficult to forecast. The first consideration is that the USSR, with its vast Arctic continental shelf and unimaginably large potential petroleum resources, starts working actively. Activity in the western world would probably then be advanced. The other possibility is an international oil price or supply disruption. There is a real threat that either prices suddenly go up again or a shortage of supply is created by a Mid-East conflict, or other crisis.

With crises come crash programs. ''Get domestic production quickly!'' With haste comes accidents and with them pollution, high costs and inefficiencies. The result of this sudden speed-up would, in the long run, be a strong anti-development pressure. I think we, as scientists and engineers find that an unsatisfactory scenario. But, based on fundamental economics, it will be very difficult to sustain a gradual, progressive and organized cycle of development. Investors are limited in ability or willingness, to make the front-end investments that are required. This is because industry and governments are uncertain if and when that petroleum will be needed, and at what cost. Today, we are actually uncertain if large reserves are there at all. Arctic exploration is not a top priority. While we often decry the fact that governments don't put more money into Arctic infrastructure or research, we also know the many other financial requirements of governments.

What we need is a means to reduce the uncertainties. That implies fostering research investment so that we can answer some of those questions and inspire a deliberate Arctic policy.

POLICY AND COLLABORATION

It is quite obvious that there is immense Arctic expertise around this world. It is obvious, too, that we are not communicating internationally as well as we should. We share this ocean; we should share our expertise. We have talked about nation-to-nation bilateral or multi-lateral agreements. This formality isn't what I am talking about. I am talking about people like ourselves working openly and informally—talking to each other and exchanging information. This has been the major achievement of this Symposium and I congratulate the Marine Technology Society for organizing this dialogue among different sectors—government, industry and university, among different disciplines and among several nations. It is what is needed and we need to do more of it. Before we can really be effective, we have to get our houses in order. By that, I mean that in each of our countries we need to have much greater rappport among government, industry and university research centers. There has been a growing tendency of an individual to consider him or herself a member of a particular sector and to work in isolation of others. Much more can and needs to be done to avoid redundancy and to increase our knowledge and perspective of the Arctic. Even within government, within industry and within university there exists isolation. The petroleum industry, for instance, although well coordinated oil company to oil company, is characterized by contractors and firms that don't participate sufficiently in the research and planning.

The principal message that comes from our meeting here together is that if we are going to get support from our governments, employers and the public it will be because we are organized internally within our sectors and good domestic and international dialogue is established. This presents the greatest potential for really attacking our problems in the future—the harnessing of all our talents.

When we talk about policy, generally we are speaking about government policy. We need government policy to give some direction, commitment and fiscal treatment. Wouldn't it be delightful if somebody would say, "we are going to make it possible to develop offshore oil and gas resources from the Arctic and do all the work that is necessary beforehand to make sure it is done correctly." Yes it would, but who is convinced today that the Arctic offshore merits such high national priority?

In addition to policy, we need to look to government to provide many of the infrastructural facilities. When a commercial oil field is found in the Arctic offshore, it will contain a great deal of oil, but probably will be only a few square miles in area. Compared to the vast Arctic, such an oil field would be the size of a grain of sand in this room. The Arctic needs greater communications, navigation and transportation facilities. We look to government to supply them. Government needs to fund some of the research and it is probably going to be necessary for government to give tax incentives to industry to maintain its investments.

Equally important, policy needs to be enunciated by non-government organizations. A prime example is the matter of collaboration—fostering joint research programs, gathering generic data, sharing it widely and releasing it openly, and participating in the kind of discussions we are enjoying here in Williamsburg. Just a comment on these kinds of discussions: During the 25 years I have been involved in the Arctic, I have attended perhaps 100 conferences on the topic. A frustration I share with others is that a very large percent of what is said I had heard before, some of it five or ten years previously. The reason is, that conferences invariably start from "square zero." About the third day they sometimes reach the "cutting edge," and are ready to take a bold step forward, just as the conference ends. Can we not find some way to start our conferences with the basics behind us?

Another important policy from non-government organizations is to do a much better job in informing the public about what we are doing. The public's knowledge of what is being done in preparation for responsible development of the Arctic is extremely limited. We all need to publicize our research and achievements. Above all, we need to demonstrate that we recognize what the problems are, if we are to attenuate one of the constraints to Arctic development that I mentioned earlier.

ROLE OF PROFESSIONAL SOCIETIES

How are we going to avoid the ''crash'' scenario? How are we going to remove the uncertainties.? Where is the leadership? These are important questions and I think we should not look to governments to provide the answers. I submit that it is the professional societies, such as the Marine Technology Society, that must provide leadership and members that can collectively develop policy and collaborate. The proceedings of this Symposium could serve as a document establishing a baseline from which we can more forward. The strength of international professional societies is that the membership encompasses all sectors. If they are convinced that it is important to develop Arctic offshore resources, and believe it is in the national and public interest, as well as their own, they will see that it takes place.

This Symposium has been quite successful, and I take this opportunity to say how grateful we all are to the organizers for having the vision to bring us together to talk about Arctic engineering in the 21st century. I think it is an example, even a model, of a step towards coming to grips with our problems and I certainly encourage you to go the next step.

Discussion

MR. VANCE: I wholeheartedly agree with what Ernie said. I have been in the business for over 25 years, but I have heard the same things for 25 years. Policy in this nation is ''let nature take its course,'' as far as the Arctic is concerned, and we don't have economic or strategic reasons that the Russians have to be up there.

The only solution that I see to the problems we have tabled here is to form what I would call a North American Arctic Research Institute that would have the blessing of government. We, in the oil industry, have constraints on us. For instance, antitrust laws prevent us from cooperating in depth in the research area. Whenever my company gets together with another company, the lawyers have to be very careful about antitrust. There are many things we do that overlap. We should take that money where it overlaps and put it into an Arctic Research Institute, made up of industry, academia and government and having the ability, capability and the charter to do this forward research unhindered by antitrust laws. Then we may be getting someplace, but the policy today—the antriturst, the competition between the companies to get there first with the best technology—prevents this coordinated approach. I don't know how such institutes are formed, but there are certain precedents for it. Recently Admiral Inman formed an electronics firm in Austin to coordinate research activities in the electronics industry. Why not coordinate research in the Arctic? I don't know how it would come about. I don't know what people are needed to make it come about, but that is the only way I can see that we will have this coordinated North American Arctic Research effort.

DR. STEWART: Thank you. I agree. Is there anyone else? Yes?

MR. PARRISH: I have been sitting at this conference for three days and, like everyone else, holding in my hand a copy of the Arctic Research and Policy Act of 1984, which President Reagan signed on July 31. Congress had been working on this measure for about three or four years. It makes several different findings relating to research and the defense aspects of the Arctic and also establishes an Arctic Research

Commission. I don't know whether you are acquainted with this Act or not.

MR. VANCE: Not technically. I knew that it was being formulated and passed, but I didn't know exactly what the Commission's responsibilities were with regard to it.

MR. PARRISH: I had no preparation for doing this, but I think it is appropriate to read what the drafters of the bill said were their intentions. They are to establish a national policy, priorities and goals and to provide a federal program plan for basic and applied scientific research with respect to the Arctic, including natural resources and materials, the physical, biological and health sciences and the social and behavioral sciences. Another major purpose is to establish an Arctic Research Commission to promote Arctic research and to recommend Arctic research policy. Another is to designate the National Science Foundation as the lead agency responsible for implementing Arctic research policy, and a final purpose is to establish an Arctic Research Policy Committee to develop a national Arctic research policy and a five-year plan to implement that policy.

Now, you know, there is room for a lot of writing and a lot of research on this, and I see some potential conflict between the Arctic Research Commission and the Arctic Research Policy Committee. I suppose that will be resolved in time or perhaps by agreement between the two parties, but anyway this document is something I suggest everyone procure.

DR. PALLISTER: I am familiar with that document and have also been involved with a subcommittee studying the implementation of the President's National Security Directive No. 90. through Bob Friedheim at USC, who has an article in the Marine Technology Society *Journal* (Vol. 18, No. 1) which was included in our materials at this Symposium. You will wish to read that. He describes some of the key problems and the international implications. I could add that one of the first steps was to involve U.S neighbors and that Canadians are involved in the planning process. I am very happy that these statements by politicians have been made. That is a wonderful first step. Translating them into expenditure on research that is going to lead to responsible research development is a long process, and when they say, "Arctic research," that is very broad.

So, while pleased, I am concerned that policies not be developed in isolation. Perhaps this Society, via this Symposium, should contribute by saying, "Yes, we support these initiatives. And, here is what we are prepared to do to help out." That is when action will take place. I really stress the importance of involvement by professional societies because, by definition, the membership is open to everyone irrespective of sector and the membership is international. It is a good starting point. The weakness of professional societies is that they are sometimes underfunded, and cannot spend money being "organizers." Maybe a transition is needed. Some professional societies, such as in the medical and engineering fields do get fully active in promoting the mutual interests of society and its members.

General Discussion

MR. GERWICK: We have now reached that period for which we have approximately one-half hour for general discussion in order to address some of the overall aspects of this conference. Two or three of you have approached me and asked to speak. Perhaps Dr. Armstrong would like to speak first.

DR. ARMSTRONG: A very quick point, Mr. Chairman, and a general one. It has been mentioned two or three times that it would be desirable to have an Arctic center for research, for information, for all sorts of purposes, and I would simply like to emphasize that I suppose in the western world today there are about 30 centers of this kind at present existing—polar, Arctic, Antarctic, with different emphases, and so on.

Now, not all these, of course, are addressing problems that people here would like to see answered, but on the other hand, some of them are, and some of them are multidisciplinary places where a great many different aspects of the polar scene are studied, and there would be something relevant, I think, in what some of those had stored up that would be useful here. So, before another center is established, I think it would be worth people's while to look along this already quite well stocked shelf and see what is there, because the whole thing about these centers is that they cannot be built up quickly. An injection of a massive sum of money in the hope that that will replace time just doesn't work. So, it could be a better bet to build on something or perhaps to encourage two or three existing centers to join forces, rather than to begin again from scratch.

MR. GERWICK: We have some other general discussion. Mr. Geer?

MR. GEER: The point I would like to make has to do with the frustrations expressed in some of the discussions with regard to having heard a number of the same things presented in this Symposium that have occurred many times over the last 20-plus years. Dr. Pallister commented that he seemed to have been there before in his 25 years of involvement in the Arctic. George Vance commented that he had had similar experiences, and I am another person who has experienced some of those same frustrations, along with Ben Gerwick and a number of my other colleagues from the petroleum industry. Many of us have participated in a great number of these types of workshops. Personally, I have participated in every major workshop of this nature for the last 15 years, including the Stratton Commission report in 1969, enunciating many, if not all of the same kinds of issues that have been addressed in this Symposium.

Presently there is a study going on by OTA, reassessing a number of the same sorts of things that we addressed here today in terms of the state of technology and its ability to cope in an economical way with developing the deeper waters of the continental margin of the United States, namely, the exclusive economic zone, as well as the Arctic. MMS conducted a workshop in November one year ago addressing many, if not all the same issues. A number of recommendations were then made. The point made this morning with regard to needing continuity and "stick-to-itness," with continuity and planning over multiple years, is a critical issue.

My point here is to look back on what has happened in the past to see if there isn't a better mechanism to determine what the benefits of this Symposium might be.

So my question really is what is to be the result of this Symposium, not withstanding the fact it will be presented to all of us in a timely fashion in a nicely tied up editorial package, but what will its message be, and how will it be presented to people who can cause something to happen?

I have a statement I would like to read here, and then I would ask for any comments. The statement goes as follows: ''There are two roads that lead to an important goal and to the doing of great things. They are strength and perseverance. Strength is the lot of a few privileged men, but austere perseverance may be employed by the smallest of us and rarely fails if its purpose or its silent power grows irrestibly greater with time.''

What I would like to propose is that we employ the strength and perseverance of a great man, Dr. Spilhaus. If you are going to have a message out of this Symposium, it needs to be delivered by one who can have some meaningful impact. As I told Ben earlier, I believe in consistency in propaganda, if you want to call it that, but if we tell the story, then there ought to be built in it some strength from previous workshops.

My only point here now is what should be done to accommodate that in terms of the plans MTS might have to cause this to happen and to find out the nature of the audience to whom this should be addressed. I leave that to your good thoughts. I don't have any ideas or suggestions on what that mechanism might be, but it is only to point out that many of these issues, not all of them, but many of these issues have been dealt with in great detail in the past, and unfortunately, nothing really concrete has occurred from the tremendous amount of mental effort by great numbers of dedicated people. In response to Jerry Machemehl's comment earlier about not having greater presentation from the petroleum industry at this particular Symposium, it simply stems from the fact that the subject matter has been worked over and over again, and the question is how many times can one work this subject and benefit from it? Walt Spring with Mobil and I have recently participated in a number of these things, some of which were considered illegal by some current Congressional people. The question I have is what is going to be the mechanism by which something beneficial can occur from such an intellectual exchange as has taken place at this Symposium?

MR. GERWICK: You have given us quite a challenge on a late afternoon.

Do I have any other discussion from the floor? Yes, please, Dr. Carstens?

DR. CARSTENS: Mr. Chairman, I think Mr. Geer just disclosed the excellent strategy of MTS in naming this Symposium after Dr. Spilhaus. This way it may have more of an impact than it would otherwise have.

The fact that I grew up way beyond the Arctic Circle may have something to do with what I am going to say to you now. It has to do with our attitude towards ice. It is true, of course, that ice is our biggest problem. It is our enemy. But it is also true that ice is solving many problems for us. So, ice is our friend, as well. A good example may be the well-known case from the Beaufort Sea Ice Islands. They suffer in the summer from waves, but as soon as the wave season ends with an ice cover, the troubles are over. So, the ice is our friend. Another very good example is the efficient use of ice platforms for exploratory drilling in Canada. If we allow ourselves to think constructively about how to take advantage of the ice, I have no doubt that we shall come up with a lot of uses that make good sense, both from an economic and from a safety point of view.

Remember we have plenty of time to think before the beginning, not to mention the end of the next century. To resist ice with ice, how do we go about that? We can improve and strengthen property by adding aggregate fibers. We can improve the flexural and shear strength. We can reduce creep. We can control density. So, we can produce an ice-crete much as we now produce concrete, with just about any desired properties, as long as we make sure to keep the ice cold. And we can produce ice composites reinforced with fibers, geotextiles, etc., to take care of many needs.

Probably there is quite a future for ice in hybrid structures where ice can replace other materials in footings, berms and other substructures. This would be only a very slight extrapolation of the present-day techniques of using ice or permafrost for bridges, highways, airstrips and so on. For almost any structure there is an alternative using ice. I guess it is too much to ask this audience to become an ice fan club, but I would like to read in the ensuing proceedings a recommendation to explore the potentials of ice as a material for structures of many types.

I realize that massive vested interests in conventional technology tend to be powerful deterrents to any new ideas or new technology. This is nothing specific to Arctic engineering. If anything, I think we may be better off and chances of succeeding with new ideas are better with us because we have just started, and we are not so firmly committed one way or the other. I think we will do our job best if we keep our minds open. The point that I want to make is that we should abandon the cold war against the ice and replace it with some clever ice manipulation and utilization combined with a limited but firm warfare when it is unavoidable.

CONCLUDING REMARKS

Mr. Gerwick

I should now like to share with you a few thoughts I have had, summing up the results of this conference. Like most of you, I have tried to see just what we have accomplished in the three days spent here. I think we have established that the technology does exist to proceed with the various work that we are doing, whether it be scientific research, the transport and supply of the Prudhoe Bay area and the Bering Sea, or the oil and gas production in the shallow waters of the southern Beaufort and Bering Seas. I think at the same time, we have identified a number of the areas that need continued work if we are to be ready for the challenges of the year 2000 and beyond.

Now, the problem is that in most of these areas, the very fact that they have been identified means that some of you are already working in those areas and therefore, we can see the potential for advances and for a more thorough knowledge about how to deal with the problems. One of the areas that I want to include is that mentioned by Dr. Carstens, which is the use of ice as structural material. Certainly ice roads and ice drilling pads have been used for quite awhile, but it is not too much, too futuristic, to predict that there will be an increased use of ice and, of course, of ice-bounded artificial soils, such as the frozen sands in the ESSO caisson-retained ring in the Gulf Mac. The area of soils, of course, continues to be one of great difficulty, one which will become increasingly difficult as we move further out into deeper water. As has been previously identified, we have the problems of the Arctic silts, the partially-bonded permafrost, shallow gas, soft Holocene deposits and weak clays. Development of methods of stabilizing these and strengthening them, can have a tremendous impact upon the economy of work out in the shear zone and beyond.

These are but two examples of many areas in need of continued work. Almost every speaker has asked for continuity—continuity of scientific research, and continuity of engineering research and development. I think perhaps even though it is not very dramatic, this might be one of the main messages of this conference—the need to continue what we are doing, to perservere, as Mr. Geer has said, with even greater effort. This immediately leads us to the need for much more in the way of cooperation and interchange—interchange between the military, to the extent that security conditions permit, and industry, government and academia. I think, personally, that more of this interchange is taking place on an informal basis than is generally recognized. It comes about, of course, as we work in joint industry groups, as we meet in meetings like this, as various individuals and companies work in one area and then in another; so there is quite an exchange of data. Anything we can do to improve this will be useful.

A major area identified was that of the need of more environmental data, and I am thinking specifically of ecological data for the long term. Now, Dr. Jahns identified the bowhead whale as one of immediate concern. I think perhaps in one way you could say that is outside the scope of this conference because we are supposed to be looking at the challenges of the 21st century, and that is certainly a challenge of today. However, in discussions with him at lunch, he said that if we don't solve the challenges of today, we will never make the 21st century. Certainly we have to address the bowhead whale, but beyond that immediate problem, I see the generic problem of gathering enough data far enough in advance so we understand the total ecological picture sufficiently well that we aren't constantly coming up against new crises.

What items didn't we address? I did not hear sufficient discussion of what we are going to do with the immense deposits of gas, a tremendous energy source. I think we put that off with the assumption that maybe in the latter part of the 21st century we would somehow drill, produce and pipe the gas to the southern states. I think we should have looked at the alternatives that are being explored elsewhere in the world, in the tropical and temperate zones; places such as Kalimantan, Saudi Arabia and Western Australia. There are the petrochemical plants, fertilizer plants, and alumina plants in which the gas is being transformed into a useful product. The potential, of course, of integrating gas production with coal deposits of the Brooks Range to produce petrochemicals in highly automated plants seems to be quite attractive. Once again, although I have been challenged by many of you on my earlier allusions to and comparisons with Saudi Arabia; once again, the idea of building a plant in the Arctic, manned only by a handful of people and largely automated, does not seem to me to be too far-fetched for the 21st century.

Now Dr. Pallister gave us two scenarios, and the first one, and the one I believe we have been thinking about, is that of a continuum. Things are going to continue to develop on an orderly basis right on through the next 20 years. I think we would all like that. I think we would feel very comfortable with that scenario. However, the other scenario he gave, that of a series of quantum jumps brought about by crises, unfortunately has been more of a common pattern, at least, during the 20th century, and I think that if we are going to be ready to handle these small and big crises effectively, economically and safely, we will have to base it on intensified and continuous engineering development and scientific research working hand in hand, carried on during the more quiescent years.

Finally, let us address the question as to what use will be made of this Symposium, specifically and directly. Before I answer that, I must say that I believe that conferences, if they are as provocative as this one has been and if people really do interact and move ahead in their thinking (as I observed in all the workshops I attended), then I think conferences do do some good in themselves by disseminating, stimulating and motivating activity. Now, back to specifics. As was announced earlier by Hal Palmer, there will be a report prepared of this Symposium. The report will be sent to all the congressmen who are involved in the oceans or the Arctic in any way and to their staffs, and it will be specifically distributed to the Office of Technology Assessment in the hope that it will serve to reinforce and augment the data and the information they are getting from other sources. Copies of this report will be sent to each of you. Copies will also be distributed to various federal and Alaskan officials. We hope the report will help guide legislative and executive decisions and also contribute to the planning.

I, for one, have been extremely pleased with the Symposium that we have had here. I would like to thank each and every one of you for attending. Many of you have come extremely long distances. I hope you found it worthwhile, both within the formal setting of the conference and, also, in your social and personal interactions.

PART V:

APPENDICES

A. ATTENDEES

B. *ARCTIC RESEARCH AND POLICY ACT OF 1984*

C. BACKGROUND ABSRACTS

D. TECHNICAL PRESENTATIONS

Appendix A:

Attendees

The Symposium drew 120 ocean engineering experts from 8 nations, with the largest representations from Canada and the United States.

Norman D. Albertsen
U.S. Naval Civil Engineering Lab.
Port Hueneme, CA 94043

Mark A. Alexander
U.S. Department of the Interior
BLM 690
1800 C Street, N.W.
Washington, DC 20240

Vera Alexander
Institute of Marine Science
University of Alaska
Fairbanks, AK 99701

Itaru Aoki
Nippon Steel USA, Inc.
345 Park Avenue
New York, NY 10154

Terence Armstrong
Scott Polar Research Institute
Cambridge, England CB2 1ER

Andrew Assur
U.S. Army CRREL
48 Rip Road
Hanover, NH 03755

Susan Bales
David Taylor NSRDC
Bethesda, MD 20084

Marius Barras
Vevey Engineering Works, Ltd.
Rue des deux Gares
CH-1800 Vevey
Switzerland

Robert H. Beskow
Chevron Corporation
Suite 1070, 3 Embarcadero
San Francisco, CA 94520

John R. Botzum
Nautilus Press
Washington, DC

William Boze
Newport News Shipbuilding
4101 Washington Avenue
Newport News, VA 23607

Max C. Brewer
U.S. Geological Survey
4200 University Avenue, Rm. 224
Anchorage, AK 99508-4667

Lawson W. Brigham
U.S. Coast Guard
Ice Operations Division
2100 2nd Street, S.W.
Washington, DC 20593

Douglas Brown
NUSC/NL
Ft. Trumbull,
New London, CT 06320

Torkild Carstens
Norwegian Hydrodynamic Laboratory
P.O. Box 4118—Valentinlyst
7001 Trondheim
Norway

Jakie Chappuis
Vevey Engineering Works Ltd.
Rue des deux Gares
Ch-1800 Vevey
Switzerland

Jack J. Clark
C-CORE
Memorial University of Newfoundland
St. Johns, Newfoundland A1B 3X5
Canada

Allen Conley
Newport News Shipbuilding
4101 Washington Avenue
Newport News, VA 23607

Emilio N. Corona
ARCO Oil and Gas Co.
P.O. Box 2819 DAB 30-056
Dallas, TX 75221

Lawrence E. Cosgriff
Bugsier Towing & Salvage
P.O. Box 1717
Baltimore, MD 21203

Tom Coughlin
Vector-Schlumberger
555 Industrial Road
Sugarland, TX 77478

John P. Crawford
ARCO Oil & Gas Co.
2300 Plano Parkway
Plano, TX 75075

Kenneth R. Croasdale
K.R. Croasdale & Associates, Ltd.
334 40th Avenue, S.W.
Calgary, Alberta, 2TS OX4
Canada

John D. Crowley
Bath Iron Works Corporation
700 Washington Street
Bath, ME 04530

James S. Cumming
U.S. Coast Guard (G-WER)
2100 2nd Street, S.W.
Washington, DC 20593

W.T. Davis
U.S. Coast Guard
600 Water Street, S.W.
Washington, DC 20024

Warren W. Denner
SAIC
205 Montecito Avenue
Monterey, CA 93940

James F. Devine
U.S. Geological Survey
National Center
Reston, VA 22091

Gerald Diaz
Harding Lawson Assoc.
1541 Parkway Loop
Tustin, CA 92680

Ira Dyer
MIT
Room 5-212
77 Massachusetts Avenue
Cambridge, MA 02139

R.L. Earp
AT&T, Inc.
P.O. Box 20046
Greensboro, NC 27420

David E. Emerson
Brown & Root International
P.O. Box 3, (03) 1005
Houston, TX 77001

Terence J. Farrell
Newport News Shipbuilding
4101 Washington Avenue
Newport News, VA 23607

W.H. Felton
Newport News Shipbuilding
4101 Washington Avenue
Newport News, VA 23607

E.P. Ferrero
U.S. Department of Energy
FE-33, MS D122
Washington, DC 20545

Warren S. Fine
B.G. Offshore Constructors
One Embarcadero Center
San Francisco, CA 94111

A.E. Fiorato
Portland Cement Association
5420 Old Orchard Road
Skokie, IL 60077

Michele Fitzpatrick
USCG
G-DMT-3/54
2100 2nd Street, S.W.
Washington, DC 20593

J. Robert Fricke
ARCO Exploration Co.
P.O. Box 2819
Dallas, TX 75221

Charles Garland
J.J. Henry Co., Inc.
Two World Trade Center
Suite 9528
New York, NY 10048

Ronald L. Geer
Shell Oil Company
P.O. Box 2463
Houston, TX 77001

Nuri T. Georges
Stone & Webster Engineering Corp.
245 Summer Street
Boston, MA 02107

Ben C. Gerwick, Jr.
500 Sansome Street
San Francisco, CA 94111

Linda K. Glover
NACOA
3300 Whitehaven Street, N.W.
Washington, DC 20235

Jean A. Geortner
NSWC/White Oak
3208 Regina Drive
Silver Spring, MD 20906

C.C. Grassia
AT&T Technologies, Inc.
P.O. Box 20046
Greensboro, NC 27420

George Gryc
U.S. Geological Survey
345 Middlefield Road, MS 144
Menlo Park, CA 94025

Robert D. Hadley
BP International
Britannic House
Moor Lane
London EC2Y 9BU
England

Richard M. Hayes
U.S. Coast Guard
2100 2nd Street, S.W.
Washington, DC 20593

Dave Humphreys
c/o Commandant (G-ENE-5A)
U.S. Coast Guard
2100 2nd Street, S.W.
Washington, DC 20593

M.E. Iorns
1512 Lakewood Drive
W. Sacramento, CA 95691

Kametaro Itoh
Nippon Steel Corporation
5-10-1 Fuchinobe,
Sagamihara 229
Japan

L.R. Jacobsen
General Dynamics
Eastern Point Road
Groton, CT 06340

Hans O. Jahns
Exxon Production Research Co.
P.O. Box 2189
Houston, TX 77001

Bruce D. Johnson
Bruce D. Johnson & Assoc.
310 Ivy Lane
Dickinson, TX 77539

Richard A. Johnson
U.S. Dept. of Interior
U.S. Geological Survey
National Center
Reston, VA 22091

Betty Jones
Analysis & Technology, Inc.
Penthouse, Suite 600
2531 Jefferson Davis Hwy.
Arlington, VA 22202

Pauli Jumppanen
Technical Research Center of Finland
Laboratory of Structural Engineering
Vuoriharjuntie 33
02340 Espoo
Finland

Takeo Kayaki
Kajima Corporation
Park Avenue Plaza
55 E. 52nd Street
New York, NY 10055

Daniel Kevin
Office of Technology Assessment
U.S. Congress
Washington, DC 20510

Judith Kildow
MIT
77 Massachusetts Avenue
Cambridge, MA 02139

Dave Koski
ABAM Engineers
500 South 336th Street
Federal Way, WA 98003

William H. Kumm, President
Arctic Energies, Ltd.
511 Heavitree Lane
Severna Park, MD 21141

Robert Lai
TRW, Inc.
7600 Colshire Drive
McLean, VA 22102

Robert J. Lantz
U.S. Geological Survey
345 Middlefield Road, MS 187
Menlo Park, CA 94025

Thomas L. Laughlin
Office of Policy and Planning
NOAA
14th & Constitution Avenue, N.W.
Washington, DC 20230

Jerry L. Machemehl
ARCO Oil & Gas Co.
P.O. Box 2819
Dallas, TX 75221

Stuart Mackiernan
General Dynamics
97 E. Howard Street
Quincy, MA 02169

Larry J. Mahar
Earth Technology Corporation
3777 Long Beach Blvd.
Long Beach, CA 90807

Eero Makinen
Wartsila Arctic, Inc.
Suite 201
1684 W. 8th Avenue
Vancouver, B.C. V6J 1V4
Canada

Charles E. Malanick
818 Harrington Road
Rockville, MD 20852

Preston Mason
Seaflo Systems
P.O. Box 42260
Houston, TX 77242

Gilbert L. Maton
Tracor Applied Sciences, Inc.
Washington Operations
1601 Research Blvd.
Rockville, MD 20850

Scott McDowell
EG&G WASC
Oceanographic Services
77 Rumford Avenue
Waltham, MA 02154

D.L. McPherson
Gulf Canada Resources, Ltd.
P.O. Box 130
Calgary, Alberta T2P 2H7
Canada

James B. Montgomery
Newport News Shipbuilding
4101 Washington Avenue
Newport News, VA 23607

George Mott
Texaco U.S.A.
P.O. Box 60252
New Orleans, LA 70160

George B. Newton
Analysis & Technology, Inc.
Penthouse, Suite 600
2531 Jefferson Davis Hwy.
Arlington, VA 22202

John M. Niedzwecki
Texas A&M University
Dept. of Civil Engineering
College Station, TX 77843

H.N. Opland
NORDA
P.O. Box 29689
New Orleans, LA 70129

Masanao Oshima
Mitsui Engineering & Shipbuilding
5-6-4 Tsukiji, Chuo-Ku
Tokyo 104
Japan

Willy Ostreng
The Fridtjof Nansen Institute
Fr. Nansensvei 17
1324 Lysaker
Norway

A. Ernest Pallister
Pallister Resource Management, Ltd.
Box 3500
Calgary, Alberta T2P 2P9
Canada

Harold D. Palmer
Marine Technology Society
2000 Florida Avenue, N.W.
Suite 500
Washington, DC 20009

J.A. Palmer
Newport News Shipbuilding
4101 Washington Avenue
Newport News, VA 23607

Max R. Parrish
Parrish & Company, Inc.
1825 Eye Street, N.W., Ste. 400
Washington, D.C. 20006

John S. Parsons
AMETEK/ORED
1224 Coast Village Circle
Santa Barbara, CA 93108

Denzil C. Pauli
13021 Bluhill Road
Silver Spring, MD 20906

John J. Peebles
Morrison-Knudsen Co.
M-K Plaza II
Boise, ID 83729

Donald W. Perkins
National Research Council
Marine Board
2101 Constitution Avenue, N.W.
Washington, DC 20418

Abe Person
Global Marine Development
P.O. Box 4577
Houston, TX 77210

Jack Pollack
IMODCO
10960 Wilshire Blvd.
Suite 1100
Los Angeles, CA 90024

Hank Porterfield
6441 Old Chesterbrook Rd.
McLean, VA 22101

Richard D. Pritchard
Sea Ice Consultants, Inc.
4710 Auth Place, Suite 175
Suitland, MD 20746

Louis Rey
Comité Arctique International
La Jaquière
Epalinges 1066
Switzerland

Frank P. Robertson
Morrison-Knudsen Co., Inc.
P.O. Box 7808
Boise, ID 83729

Michael Rojansky
Ben C. Gerwick, Inc.
500 Sansome Street
San Francisco, CA 94111

William M. Sackinger
University of Alaska
Geophysical Institute
Fairbanks, AK 99701

John F. Schindler
Department of the Interior
Minerals Management Service
2473 Captain Cook
Anchorage, AK 99503

Craig Schnappinger
U.S. Coast Guard
U.S. Coast Guard Academy
New London, CT 06320

Richard N. Seemel
Bechtel Civil & Min., Inc.
122 Galewood Circle
San Francisco, CA 94131

Richard M. Shamp
Engineering Services Associates, Inc.
1500 Massachusetts Ave., N.W.
Washington, DC 20005

Harold D. Shoemaker
U.S. Department of Energy
P.O. Box 880
Morgantown, WV 26505

Edward L. Smith
ODSI Defense Systems, Inc.
6110 Executive Blvd.
Rockville, MD 20852

D.S. Sodhi
U.S. Army CRREL
72 Lyme Road
Hanover, NH 03755-1290

C.J. Sonu
Tekmarine
37 Auburn Avenue
Sierra Madre, CA 91024

Athelstan F. Spilhaus
P.O. Box 1063
Middleburg, VA 22117

Walter Spring
Mobile R&D
13777 Midway Road
Dallas, TX 75234

Harris B. Stewart, Jr.
Director, Center for Marine Studies
Old Dominion University
Norfolk, VA 23508

Irv Tailleur
U.S. Geological Survey
345 Middlefield Road, MS 187
Menlo Park, Ca 94025

Richard Taranto
Analysis & Technology, Inc.
Penthouse, Suite 600
2531 Jefferson Davis Hwy.
Arlington, VA 22202

D.S. Tolderlund
U.S. Coast Guard Academy
New London, CT 06320

C.M. Tucker
DSTI, NDHQ
101 Col. by Drive
Ottawa, Ontario K1A OK2
Canada

George P. Vance
Mobil Oil
145 King Street West
Toronto, Ontario M5J 2L4
Canada

Norman C. Venzke
Chief, Office of Operations
U.S. Coast Guard (G-O 3112)
2100 2nd Street, S.W.
Washington, DC 20593

Richard P. Voelker
ARCTEC, Inc.
9104 Red Branch Road
Columbia, MD 21045

Gunter Weller
University of Alaska
Geophysical Institute
Fairbanks, AK 99701

James W. Winchester
NOAA
14th & Constitution Avenue, N.W.
Room 5126
Washington, D.C. 20230

Dan Womack
Parker Drilling Company
8 East 3rd
Tulsa, OK 74103

L.C. Zaleski-Zamenhof
C.G. Doris
58A, Rue du Dessous des Berges
75013 Paris
France

REGISTERED BUT DID NOT ATTEND:

J.T. Gottwald
Tracor Hydronautics
7210 Pindell School Road
Laurel, MD 20707

George W. Martin
Project Engineering, Inc.
8630 Fenton Street, Suite 115
Silver Spring, MD 20910

Alan W. Niedoroda
R.J. Brown & Associates
2010 North Loop West, #200
Houston, TX 77018

James A. Rickard
Sohio Petroleum Company
One Lincoln Centre
5400 LBJ Freeway, Suite 1200
LB/25
Dallas, TX 75240

Fred Roots
Department of Fisheries and Environment
Ottawa, Ontario K1A OH3
Canada

Einar Skinnarland
North Atlantic Contractors
11 Viamede Crescent
Willowdale, Ontario, M2K 2A7
Canada

MTS STAFF:

Mary Beth Hatem
Managing Editor
MTS Journal

Jan Vorhees
Technical Programs Coordinator

Appendix B:

Arctic Research and Policy Act of 1984

On July 31, 1984, Public Law 98-373—the ''Arctic Research and Policy Act of 1984''*—was signed into law by the 98th Congress. The primary purpose of the legislation is to provide for a comprehensive national policy for national research needs and objectives in the Arctic.*

Public Law 98-373
98th Congress

An Act

July 31, 1984
[S. 373]

To provide for a comprehensive national policy dealing with national research needs and objectives in the Arctic, for a National Critical Materials Council, for development of a continuing and comprehensive national materials policy, for programs necessary to carry out that policy, including Federal programs of advanced materials research and technology, and for innovation in basic materials industries, and for other purposes.

Be it enacted by the Senate and House of Representatives of the United States of America in Congress assembled,

Arctic Research and Policy Act of 1984.

TITLE I—ARCTIC RESEARCH AND POLICY

SHORT TITLE

15 USC 4101 note.

SEC. 101. This title may be cited as the "Arctic Research and Policy Act of 1984".

FINDINGS AND PURPOSES

15 USC 4101.

SEC. 102. (a) The Congress finds and declares that—

(1) the Arctic, onshore and offshore, contains vital energy resources that can reduce the Nation's dependence on foreign oil and improve the national balance of payments;

(2) as the Nation's only common border with the Soviet Union, the Arctic is critical to national defense;

(3) the renewable resources of the Arctic, specifically fish and other seafood, represent one of the Nation's greatest commercial assets;

(4) Arctic conditions directly affect global weather patterns and must be understood in order to promote better agricultural management throughout the United States;

(5) industrial pollution not originating in the Arctic region collects in the polar air mass, has the potential to disrupt global weather patterns, and must be controlled through international cooperation and consultation;

(6) the Arctic is a natural laboratory for research into human health and adaptation, physical and psychological, to climates of extreme cold and isolation and may provide information crucial for future defense needs;

(7) atmospheric conditions peculiar to the Arctic make the Arctic a unique testing ground for research into high latitude communications, which is likely to be crucial for future defense needs;

(8) Arctic marine technology is critical to cost-effective recovery and transportation of energy resources and to the national defense;

(9) the United States has important security, economic, and environmental interests in developing and maintaining a fleet of icebreaking vessels capable of operating effectively in the heavy ice regions of the Arctic;

(10) most Arctic-rim countries, particularly the Soviet Union, possess Arctic technologies far more advanced than those currently available in the United States;

(11) Federal Arctic research is fragmented and uncoordinated at the present time, leading to the neglect of certain areas of research and to unnecessary duplication of effort in other areas of research;

(12) improved logistical coordination and support for Arctic research and better dissemination of research data and information is necessary to increase the efficiency and utility of national Arctic research efforts;

(13) a comprehensive national policy and program plan to organize and fund currently neglected scientific research with respect to the Arctic is necessary to fulfill national objectives in Arctic research;

(14) the Federal Government, in cooperation with State and local governments, should focus its efforts on the collection and characterization of basic data related to biological, materials, geophysical, social, and behavioral phenomena in the Arctic;

(15) research into the long-range health, environmental, and social effects of development in the Arctic is necessary to mitigate the adverse consequences of that development to the land and its residents;

(16) Arctic research expands knowledge of the Arctic, which can enhance the lives of Arctic residents, increase opportunities for international cooperation among Arctic-rim countries, and facilitate the formulation of national policy for the Arctic; and

(17) the Alaskan Arctic provides an essential habitat for marine mammals, migratory waterfowl, and other forms of wildlife which are important to the Nation and which are essential to Arctic residents.

(b) The purposes of this title are—

(1) to establish national policy, priorities, and goals and to provide a Federal program plan for basic and applied scientific research with respect to the Arctic, including natural resources and materials, physical, biological and health sciences, and social and behavioral sciences;

(2) to establish an Arctic Research Commission to promote Arctic research and to recommend Arctic research policy;

(3) to designate the National Science Foundation as the lead agency responsible for implementing Arctic research policy; and

(4) to establish an Interagency Arctic Research Policy Committee to develop a national Arctic research policy and a five year plan to implement that policy.

ARCTIC RESEARCH COMMISSION — Establishment.

SEC. 103. (a) The President shall establish an Arctic Research Commission (hereafter referred to as the "Commission"). 15 USC 4102.

(b)(1) The Commission shall be composed of five members appointed by the President, with the Director of the National Science Foundation serving as a nonvoting, ex officio member. The members appointed by the President shall include—

(A) three members appointed from among individuals from academic or other research institutions with expertise in areas of research relating to the Arctic, including the physical, biological, health, environmental, social, and behavioral sciences;

(B) one member appointed from among indigenous residents of the Arctic who are representative of the needs and interests of Arctic residents and who live in areas directly affected by Arctic resource development; and

(C) one member appointed from among individuals familiar with the Arctic and representative of the needs and interests of private industry undertaking resource development in the Arctic.

(2) The President shall designate one of the appointed members of the Commission to be chairperson of the Commission.

(c)(1) Except as provided in paragraph (2) of this subsection, the term of office of each member of the Commission appointed under subsection (b)(1) shall be four years.

(2) Of the members of the Commission originally appointed under subsection (b)(1)—

(A) one shall be appointed for a term of two years;

(B) two shall be appointed for a term of three years; and

(C) two shall be appointed for a term of four years.

(3) Any vacancy occurring in the membership of the Commission shall be filled, after notice of the vacancy is published in the Federal Register, in the manner provided by the preceding provisions of this section, for the remainder of the unexpired term.

(4) A member may serve after the expiration of the member's term of office until the President appoints a successor.

(5) A member may serve consecutive terms beyond the member's original appointment.

(d)(1) Members of the Commission may be allowed travel expenses, including per diem in lieu of subsistence, as authorized by section 5703 of title 5, United States Code. A member of the Commission not presently employed for compensation shall be compensated at a rate equal to the daily equivalent of the rate for GS-16 of the General Schedule under section 5332 of title 5, United States Code, for each day the member is engaged in the actual performance of his duties as a member of the Commission, not to exceed 90 days of service each year. Except for the purposes of chapter 81 of title 5 (relating to compensation for work injuries) and chapter 171 of title 28 (relating to tort claims), a member of the Commission shall not be considered an employee of the United States for any purpose.

5 USC 8101 *et seq.*
28 USC 2671 *et seq.*

(2) The Commission shall meet at the call of its Chairman or a majority of its members.

(3) Each Federal agency referred to in section 107(b) may designate a representative to participate as an observer with the Commission. These representatives shall report to and advise the Commission on the activities relating to Arctic research of their agencies.

Public meeting

(4) The Commission shall conduct at least one public meeting in the State of Alaska annually.

DUTIES OF COMMISSION

15 USC 4103.

SEC. 104. (a) The Commission shall—

(1) develop and recommend an integrated national Arctic research policy;

(2) in cooperation with the Interagency Arctic Research Policy Committee established under section 107, assist in establishing a national Arctic research program plan to implement the Arctic research policy;

(3) facilitate cooperation between the Federal Government and State and local governments with respect to Arctic research;

(4) review Federal research programs in the Arctic and suggest improvements in coordination among programs;

(5) recommend methods to improve logistical planning and support for Arctic research as may be appropriate and in accordance with the findings and purposes of this title;

(6) suggest methods for improving efficient sharing and dissemination of data and information on the Arctic among interested public and private institutions;

(7) offer other recommendations and advice to the Interagency Committee established under section 107 as it may find appropriate; and

(8) cooperate with the Governor of the State of Alaska and with agencies and organizations of that State which the Governor may designate with respect to the formulation of Arctic research policy.

(b) Not later than January 31 of each year, the Commission shall—

(1) publish a statement of goals and objectives with respect to Arctic research to guide the Interagency Committee established under section 107 in the performance of its duties; and

(2) submit to the President and to the Congress a report describing the activities and accomplishments of the Commission during the immediately preceding fiscal year. Report.

COOPERATION WITH THE COMMISSION

SEC. 105. (a)(1) The Commission may acquire from the head of any Federal agency unclassified data, reports, and other nonproprietary information with respect to Arctic research in the possession of the agency which the Commission considers useful in the discharge of its duties. 15 USC 4104.

(2) Each agency shall cooperate with the Commission and furnish all data, reports, and other information requested by the Commission to the extent permitted by law; except that no agency need furnish any information which it is permitted to withhold under section 552 of title 5, United States Code. Confidentiality.

(b) With the consent of the appropriate agency head, the Commission may utilize the facilities and services of any Federal agency to the extent that the facilities and services are needed for the establishment and development of an Arctic research policy, upon reimbursement to be agreed upon by the Commission and the agency head and taking every feasible step to avoid duplication of effort.

(c) All Federal agencies shall consult with the Commission before undertaking major Federal actions relating to Arctic research.

ADMINISTRATION OF THE COMMISSION

SEC. 106. The Commission may— 15 USC 4105.

(1) in accordance with the civil service laws and subchapter III of chapter 53 of title 5, United States Code, appoint and fix the compensation of an Executive Director and necessary additional staff personnel, but not to exceed a total of seven compensated personnel; 5 USC 5331.

(2) procure temporary and intermittent services as authorized by section 3109 of title 5, United States Code;

(3) enter into contracts and procure supplies, services, and personal property; and

(4) enter into agreements with the General Services Administration for the procurement of necessary financial and administrative services, for which payment shall be made by reimbursement from funds of the Commission in amounts to be agreed upon by the Commission and the Administrator of the General Services Administration.

LEAD AGENCY AND INTERAGENCY ARCTIC RESEARCH POLICY COMMITTEE

15 USC 4106.

SEC. 107. (a) The National Science Foundation is designated as the lead agency responsible for implementing Arctic research policy, and the Director of the National Science Foundation shall insure that the requirements of section 108 are fulfilled.

Establishment.

(b)(1) The President shall establish an Interagency Arctic Research Policy Committee (hereinafter referred to as the "Interagency Committee").

(2) The Interagency Committee shall be composed of representatives of the following Federal agencies or offices:

(A) the National Science Foundation;
(B) the Department of Commerce;
(C) the Department of Defense;
(D) the Department of Energy;
(E) the Department of the Interior;
(F) the Department of State;
(G) the Department of Transportation;
(H) the Department of Health and Human Services;
(I) the National Aeronautics and Space Administration;
(J) the Environmental Protection Agency; and
(K) any other agency or office deemed appropriate.

(3) The representative of the National Science Foundation shall serve as the Chairperson of the Interagency Committee.

DUTIES OF THE INTERAGENCY COMMITTEE

15 USC 4107.

SEC. 108. (a) The Interagency Committee shall—

(1) survey Arctic research conducted by Federal, State, and local agencies, universities, and other public and private institutions to help determine priorities for future Arctic research, including natural resources and materials, physical and biological sciences, and social and behavioral sciences;

(2) work with the Commission to develop and establish an integrated national Arctic research policy that will guide Federal agencies in developing and implementing their research programs in the Arctic;

(3) consult with the Commission on—

(A) the development of the national Arctic research policy and the 5-year plan implementing the policy;

(B) Arctic research programs of Federal agencies;

(C) recommendations of the Commission on future Arctic research; and

(D) guidelines for Federal agencies for awarding and administering Arctic research grants;

(4) develop a 5-year plan to implement the national policy, as provided for in section 109;

(5) provide the necessary coordination, data, and assistance for the preparation of a single integrated, coherent, and multiagency budget request for Arctic research as provided for in section 110;

(6) facilitate cooperation between the Federal Government and State and local governments in Arctic research, and recommend the undertaking of neglected areas of research in accordance with the findings and purposes of this title;

(7) coordinate and promote cooperative Arctic scientific research programs with other nations, subject to the foreign policy guidance of the Secretary of State;

(8) cooperate with the Governor of the State of Alaska in fulfilling its responsibilities under this title;

(9) promote Federal interagency coordination of all Arctic research activities, including—

(A) logistical planning and coordination; and

(B) the sharing of data and information associated with Arctic research, subject to section 552 of title 5, United States Code; and

(10) provide public notice of its meetings and an opportunity for the public to participate in the development and implementation of national Arctic research policy. Public information.

(b) Not later than January 31, 1986, and biennially thereafter, the Interagency Committee shall submit to the Congress through the President, a brief, concise report containing— Report.

(1) a statement of the activities and accomplishments of the Interagency Committee since its last report; and

(2) a description of the activities of the Commission, detailing with particularity the recommendations of the Commission with respect to Federal activities in Arctic research.

5-YEAR ARCTIC RESEARCH PLAN

SEC. 109. (a) The Interagency Committee, in consultation with the Commission, the Governor of the State of Alaska, the residents of the Arctic, the private sector, and public interest groups, shall prepare a comprehensive 5-year program plan (hereinafter referred to as the "Plan") for the overall Federal effort in Arctic research. The Plan shall be prepared and submitted to the President for transmittal to the Congress within one year after the enactment of this Act and shall be revised biennially thereafter. 15 USC 4108.

(b) The Plan shall contain but need not be limited to the following elements:

(1) an assessment of national needs and problems regarding the Arctic and the research necessary to address those needs or problems;

(2) a statement of the goals and objectives of the Interagency Committee for national Arctic research;

(3) a detailed listing of all existing Federal programs relating to Arctic research, including the existing goals, funding levels for each of the 5 following fiscal years, and the funds currently being expended to conduct the programs;

(4) recommendations for necessary program changes and other proposals to meet the requirements of the policy and goals

as set forth by the Commission and in the Plan as currently in effect; and

(5) a description of the actions taken by the Interagency Committee to coordinate the budget review process in order to ensure interagency coordination and cooperation in (A) carrying out Federal Arctic research programs, and (B) eliminating unnecessary duplication of effort among these programs.

COORDINATION AND REVIEW OF BUDGET REQUESTS

15 USC 4109.

SEC. 110. (a) The Office of Science and Technology Policy shall—

(1) review all agency and department budget requests related to the Arctic transmitted pursuant to section 108(a)(5), in accordance with the national Arctic research policy and the 5-year program under section 108(a)(2) and section 109, respectively; and

(2) consult closely with the Interagency Committee and the Commission to guide the Office of Science and Technology Policy's efforts.

(b)(1) The Office of Management and Budget shall consider all Federal agency requests for research related to the Arctic as one integrated, coherent, and multiagency request which shall be reviewed by the Office of Management and Budget prior to submission of the President's annual budget request for its adherence to the Plan. The Commission shall, after submission of the President's annual budget request, review the request and report to Congress on adherence to the Plan.

Report.

(2) The Office of Management and Budget shall seek to facilitate planning for the design, procurement, maintenance, deployment, and operations of icebreakers needed to provide a platform for Arctic research by allocating all funds necessary to support icebreaking operations, except for recurring incremental costs associated with specific projects, to the Coast Guard.

AUTHORIZATION OF APPROPRIATIONS; NEW SPENDING AUTHORITY

15 USC 4110.

SEC. 111. (a) There are authorized to be appropriated such sums as may be necessary for carrying out this title.

2 USC 651.

(b) Any new spending authority (within the meaning of section 401 of the Congressional Budget Act of 1974) which is provided under this title shall be effective for any fiscal year only to such extent or in such amounts as may be provided in appropriation Acts.

DEFINITION

15 USC 4111.

SEC. 112. As used in this title, the term "Arctic" means all United States and foreign territory north of the Arctic Circle and all United States territory north and west of the boundary formed by the Porcupine, Yukon, and Kuskokwim Rivers; all contiguous seas, including the Arctic Ocean and the Beaufort, Bering, and Chukchi Seas; and the Aleutian chain.

National Critical Materials Act of 1984.

TITLE II—NATIONAL CRITICAL MATERIALS ACT OF 1984

SHORT TITLE

30 USC 1801 note.

SEC. 201. This title may be cited as the "National Critical Materials Act of 1984".

FINDINGS AND PURPOSES

SEC. 202. (a) The Congress finds that— 30 USC 1801.

(1) the availability of adequate supplies of strategic and critical industrial minerals and materials continues to be essential for national security, economic well-being, and industrial production;

(2) the United States is increasingly dependent on foreign sources of materials and vulnerable to supply interruption in the case of many of those minerals and materials essential to the Nation's defense and economic well-being;

(3) together with increasing import dependence, the Nation's industrial base, including the capacity to process minerals and materials, is deteriorating—both in terms of facilities and in terms of a trained labor force;

(4) research, development, and technological innovation, especially related to improved materials and new processing technologies, are important factors which affect our long-term capability for economic competitiveness, as well as for adjustment to interruptions in supply of critical minerals and materials;

(5) while other nations have developed and implemented specific long-term research and technology programs to develop high-performance materials, no such policy and program evolution has occurred in the United States;

(6) establishing critical materials reserves, by both the public and private sectors and with proper organization and management, represents one means of responding to the genuine risks to our economy and national defense from dependency on foreign sources;

(7) there exists no single Federal entity with the authority and responsibility for establishing critical materials policy and for coordinating and implementing that policy; and

(8) the importance of materials to national goals requires an organizational means for establishing responsibilities for materials programs and for the coordination, within and at a suitably high level of the Executive Office of the President, with other existing policies within the Federal Government.

(b) It is the purpose of this title—

(1) to establish a National Critical Materials Council under and reporting to the Executive Office of the President which shall—

(A) establish responsibilities for and provide for necessary coordination of critical materials policies, including all facets of research and technology, among the various agencies and departments of the Federal Government, and make recommendations for the implementation of such policies;

(B) bring to the attention of the President, the Congress, and the general public such materials issues and concerns, including research and development, as are deemed critical to the economic and strategic health of the Nation; and

(C) ensure adequate and continuing consultation with the private sector concerning critical materials, materials research and development, use of materials, Federal materials policies, and related matters;

(2) to establish a national Federal program for advanced materials research and technology, including basic phenomena through processing and manufacturing technology; and

(3) to stimulate innovation and technology utilization in basic as well as advanced materials industries.

ESTABLISHMENT OF THE NATIONAL CRITICAL MATERIALS COUNCIL

30 USC 1802. SEC. 203. There is hereby established a National Critical Materials Council (hereinafter referred to as the "Council") under and reporting to the Executive Office of the President. The Council shall be composed of three members who shall be appointed by the President and who shall serve at the pleasure of the President. Members so appointed who are not already Senate-confirmed officers of the Government shall be appointed by and with the advice and consent of the Senate. The President shall designate one of the members to serve as Chairman. Each member shall be a person who, as a result of training, experience, and achievement, is qualified to carry out the duties and functions of the Council, with particular emphasis placed on fields relating to materials policy or materials science and engineering. In addition, at least one of the members shall have a background in and understanding of environmentally related issues.

RESPONSIBILITIES AND AUTHORITIES OF THE COUNCIL

30 USC 1803. SEC. 204. (a) It shall be the primary responsibility of the Council—

(1) to assist and advise the President in establishing coherent national materials policies consistent with other Federal policies, and making recommendations necessary to implement such policies;

(2) to assist in establishing responsibilities for, and to coordinate, Federal materials-related policies, programs, and research and technology activities, as well as recommending to the Office of Management and Budget budget priorities for materials activities in each of the Federal departments and agencies;

(3) to review and appraise the various programs and activities of the Federal Government in accordance with the policy and directions given in the National Materials and Minerals Policy, Research and Development Act of 1980 (30 U.S.C. 1601), and to determine the extent to which such programs and activities are contributing to the achievement of such policy and directions;

(4) to monitor and evaluate the critical materials needs of basic and advanced technology industries and the Government, including the critical materials research and development needs of the private and public sectors;

(5) to advise the President of mineral and material trends, both domestic and foreign, the implications thereof for the United States and world economies and the national security, and the probable effects of such trends on domestic industries;

(6) to assess through consultation with the materials academic community the adequacy and quality of materials-related educational institutions and the supply of materials scientists and engineers;

(7) to make or furnish such studies, analyses, reports, and recommendations with respect to matters of materials-related policy and legislation as the President may request;

Report. (8)(A) to prepare a report providing a domestic inventory of critical materials with projections on the prospective needs of Government and industry for these materials, including a long-range assessment, prepared in conjunction with the Office of

Science and Technology Policy in accordance with the National Materials and Minerals Policy, Research and Development Act of 1980, and in conjunction with such other Government departments or agencies as may be considered necessary, of the prospective major critical materials problems which the United States is likely to confront in the immediate years ahead and providing advice as to how these problems may best be addressed, with the first such report being due on April 1, 1985, and (B) review and update such report and assessment as appropriate and report thereon to the Congress at least biennially; and 30 USC 1601 note.

(9) to recommend to the Congress such changes in current policies, activities, and regulations of the Federal Government, and such legislation, as may be considered necessary to carry out the intent of this title and the National Materials and Minerals Policy, Research and Development Act of 1980.

(b) In carrying out its responsibilities under this section the Council shall have the authority— Advisory panels and interagency committees.

(1) to establish such special advisory panels as it considers necessary, with each such panel consisting of representatives of industry, academia, and other members of the private sector, not to exceed ten members, and being limited in scope of subject and duration; and

(2) to establish and convene such Federal interagency committees as it considers necessary in carrying out the intent of this title.

(c) In seeking to achieve the goals of this title and related Acts, the Council and other Federal departments and agencies with responsibilities or jurisdiction related to materials or materials policy, including the National Security Council, the Council on Environmental Quality, the Office of Management and Budget, and the Office of Science and Technology Policy, shall work collaboratively and in close cooperation.

PROGRAM AND POLICY FOR ADVANCED MATERIALS RESEARCH AND TECHNOLOGY

SEC. 205. (a) In addition to the responsibilities described in section 204, the Council shall be responsible for coordination with appropriate agencies and departments of the Federal Government relative to Federal materials research and development policies and programs. Such policies and programs shall be consistent with the policies and goals described in the National Materials and Minerals Policy, Research and Development Act of 1980. In carrying out this responsibility the Council shall— 30 USC 1804.

(1)(A) establish a national Federal program plan for advanced materials research and development, recommend the designation of the key responsibilities for carrying out such research, and to provide for coordination of this plan with the Office of Science and Technology Policy, the Office of Management and Budget, and such other Federal offices and agencies as may be deemed appropriate, and (B) annually review such plan and report thereon to the Congress;

(2) review annually the materials research, development, and technology authorization requests and budgets of all Federal agencies and departments; and in this activity the Council shall make recommendations, in cooperation with the Office of Sci-

ence and Technology Policy, the Office of Management and Budget, and all other Federal offices and agencies deemed appropriate, to ensure close coordination of the goals and directions of such programs with the policies determined by the Council; and

(3) assist the Office of Science and Technology Policy in the preparation of such long-range materials assessments and reports as may be required by the National Materials and Minerals Policy, Research and Development Act of 1980, and assist other Federal entities in the preparation of analyses and reporting relating to critical and advanced materials.

30 USC 1601 note.

(b) The Office of Management and Budget, in reviewing the materials research, development, and technology authorization requests of the various Federal departments and agencies for any fiscal year, and the recommendations of the Council, shall consider all of such requests and recommendations as an integrated, coherent, multiagency request which shall be reviewed by the Office of Management and Budget for its adherence to the national Federal materials program plan in effect for such fiscal year under subsection (a).

INNOVATION IN BASIC AND ADVANCED MATERIALS INDUSTRIES

Centers for Industrial Technology. 30 USC 1805.

SEC. 206. (a)(1) In order to promote the use of more cost-effective, advanced technology and other means of providing for innovation and increased productivity within the basic and advanced materials industries, the Council shall evaluate and make recommendations regarding the establishment of Centers for Industrial Technology as provided in Public Law 96-480 (15 U.S.C. 3705).

(2) The activities of such Centers shall focus on, but not be limited to, the following generic materials areas: corrosion; welding and joining of materials; advanced processing and fabrication technologies; microfabrication; and fracture and fatigue.

(b) In order to promote better use and innovation of materials in design for improved safety or efficiency, the Council shall establish in cooperation with the appropriate Federal agencies and private industry, an effective mechanism for disseminating materials property data in an efficient and timely manner. In carrying out this responsibility, the Council shall consider, where appropriate, the establishment of a computerized system taking into account, to the maximum extent practicable, existing available resources.

COMPENSATION OF MEMBERS AND REIMBURSEMENTS

30 USC 1806.

SEC. 207. (a) The Chairman of the Council, if not otherwise a paid officer or employee of the Federal Government, shall be paid at the rate not to exceed the rate of basic pay provided for level II of the Executive Schedule. The other members of the Council, if not otherwise paid officers or employees of the Federal Government, shall be paid at a per diem rate comparable to the rate not to exceed the rate of basic pay provided for level III of the Executive Schedule.

(b) Subject to existing law and regulations governing conflicts of interest, the Council may accept reimbursement from any private nonprofit organization or from any department, agency, or instrumentality of the Federal Government, or from any State or local government, for reasonable travel expenses incurred by any member or employee of the Council in connection with such member's or employee's attendance at any conference, seminar, or similar meeting.

POSITION AND AUTHORITIES OF EXECUTIVE DIRECTOR

SEC. 208. (a) There shall be an Executive Director (hereinafter referred to as the "Director"), who shall be chief administrator of the Council. The Director shall be appointed by the Council full time and shall be paid at the rate not to exceed the rate of basic pay provided for level III of the Executive Schedule. 30 USC 1807.

(b) The Director is authorized—

(1) to employ such personnel as may be necessary for the Council to carry out its duties and functions under this title, but not to exceed twelve compensated employees;

(2) to obtain the services of experts and consultants in accordance with the provisions of section 3109 of title 5, United States Code; and

(3) to develop, subject to approval by the Council, rules and regulations necessary to carry out the purposes of this title.

(c) In exercising his responsibilities and duties under this title, the Director—

(1) may consult with representatives of academia, industry, labor, State and local governments, and other groups; and

(2) shall utilize to the fullest extent possible the services, facilities, and information (including statistical information) of public and private agencies, organizations, and individuals.

(d) Notwithstanding section 367(b) of the Revised Statutes (31 U.S.C. 665(b)), the Council may utilize voluntary and uncompensated labor and services in carrying out its duties and functions. 31 USC 1342.

RESPONSIBILITIES AND DUTIES OF THE DIRECTOR

SEC. 209. In carrying out his functions the Director shall assist and advise the Council on policies and programs of the Federal Government affecting critical and advanced materials by— 30 USC 1808.

(1) providing the professional and administrative staff and support for the Council;

(2) assisting the Federal agencies and departments in appraising the effectiveness of existing and proposed facilities, programs, policies, and activities of the Federal Government, including research and development, which affect critical materials availability and needs;

(3) cataloging, as fully as possible, research and development activities of the Government, private industry, and public and private institutions; and

(4) initiating Government and private studies and analyses, including those to be conducted by or under the auspices of the Council, designed to advance knowledge of critical or advanced materials issues and develop alternative proposals, including research and development, to resolve national critical materials problems.

AUTHORITY

SEC. 210. The Council is authorized— 30 USC 1809.

(1) to establish such internal rules and regulations as may be necessary for its operation; Regulations.

Contracts with U.S.

(2) to enter into contracts and acquire materials and supplies necessary for its operation to such extent or in such amounts as are provided for in appropriation Acts;

Public information.

(3) to publish, consistent with title 44 of the United States Code, or arrange to publish critical materials information that it deems to be useful to the public and private industry to the extent that such publication is consistent with the national defense and economic interest;

(4) to utilize such services or personnel as may be provided to the Council on a reimbursable basis by any agency of the United States; and

(5) to exercise such authorities as may be necessary and incidental to carrying out its responsibilities and duties under this title.

AUTHORIZATION OF APPROPRIATIONS

30 USC 1810.

SEC. 211. There are hereby authorized to be appropriated to carry out the provisions of this title a sum not to exceed $500,000 for the fiscal year ending September 30, 1985, and such sums as may be necessary thereafter: *Provided*, That the authority provided for in this title shall expire on September 30, 1990, unless otherwise authorized by Congress.

DEFINITION

30 USC 1811.

SEC. 212. As used in this title, the term "materials" has the meaning given it by section 2(b) of the National Materials and

30 USC 1601.

Minerals Policy, Research and Development Act of 1980.

Approved July 31, 1984.

LEGISLATIVE HISTORY—S. 373:

HOUSE REPORTS No. 98-593, Pt. 1 (Comm. on Science and Technology) and Pt. 2 (Comm. on Merchant Marine and Fisheries).
SENATE REPORT No. 98-159 (Comm. on Governmental Affairs).
CONGRESSIONAL RECORD:
Vol. 129 (1983): June 27, considered and passed Senate.
Vol. 130 (1984): Apr. 24, considered and passed House, amended.
June 21, Senate concurred in House amendments with amendment.
June 26, House concurred in Senate amendments.

Appendix C:
Background Abstracts

Abstracts were voluntarily submitted in advance of the Symposium and distributed as part of the participants' registration packets.

Future Developments in the Soviet Arctic Marine Transportation System

Lawson W. Brigham
U.S. Coast Guard

Arctic Work Systems Workshop

Any discussion on the future marine transportation developments in the Soviet Arctic must first reflect on the extraordinary achievements the Soviet Union has made during the past thirty years. Operation of the world's most powerful icebreakers, development of nuclear power for icebreakers, transit by ARKTIKA to the North Pole in 1977, utilization of specialized ships (shallow draft polar and river icebreakers), extension of the navigation season along the entire length of the Northern Sea Route, and routine operations with icebreaking cargo vessels, are a few, unique accomplishments of significance of the marine world.

This review will speculate on potential technological developments in the Soviet Arctic that the West may see by the next century. Replacement of a significant number of Soviet icebreakers will be necessary by the years 2000–2010. Application of nuclear power for shallow-draft polar icebreakers and lighter-cargo ships will become the norm. Independent operations of both diesel- and nuclear-powered cargo ships should become a reality. New developments in air-cushion technology, submarine barges and semi-submersibles may provide a significantly enhanced all-weather capability. Hydrocarbon development will require innovative designs for icebreaker dredges, drill ships and rescue/safety vehicles. If year-round navigation is to become a reality, extensive development of the operations infrastructure of the Soviet Arctic will be required.

ARCTIC FIXED PLATFORMS FOR WATER DEPTHS GREATER THAN 50 METERS

K. R. Croasdale
K. R. Croasdale and Associates, Ltd.

Artic Oil and Gas Workshop

The main reason why offshore platforms are more difficult to design and install in the Arctic than in other regions is because of ice. Ice creates two fundamental problems: large lateral forces, and a very short open-water season for construction.

A variety of structural configurations have been proposed for Arctic use; some are still concepts, others have been used in water depths out to 30m for exploratory drilling. Most structures built to date have been based on artificial island technology, including underwater berms on which mobile caisson structures have been placed. This type of system is good at resisting the ice and could be extrapolated for use in the 50m water depth and beyond. The economics of such systems may, however, be unfavorable because of the large amounts of granular fill required, associated with the shorter construction periods further offshore.

Alternatives to island/berm systems are steel or concrete structures designed to sit directly on the floor. The need to cope with ice has led to a variety of different forms aimed at minimizing ice forces and maximizing lateral resistance. In the author's opinion there are no technical reasons why fixed platforms of this type cannot be used beyond the 50m water depth. There is no specific water depth beyond which design ice conditions get dramatically worse (for year-round production structures). But, overturning movements increase, open-water periods get shorter for construction, and foundation conditions may get worse depending on local surficial geology.

The economics of fixed platforms beyond 50m will be very dependent on the ice loads. Recent new theories for ice forces exerted by moving pack ice indicate they may be less than previously thought. Similarly, large-scale field experiments and measurements indicate lower ice forces. This trend augurs well for using fixed platforms in the deeper Arctic waters.

In this paper a brief review of Arctic coastal ice conditions contrasting typical ice interaction scenarios for platforms in shallow and deep water is made. Platform configurations currently in use (and planned) are mentioned, and their potential for extrapolation to deeper waters is discussed. Recent developments in ice mechanics and ice load predictions are presented, together with future research needs. Finally some possible concepts for the deeper Arctic waters are presented.

Viewpoints of the Use of Fixed Platform Structures in Different Ice Conditions of The Arctic

Pauli Jumppanen
Technical Research Centre of Finland

Arctic Oil and Gas Workshop

Dredged artificial islands, caisson-retained islands and some caisson-type platforms have already been proven in shallow water conditions in the United States and Canadian Beaufort Sea. Several further structural concepts have been proposed for water depths of up to 30-40 m. In spite of extensive development of subsea working and production systems, exploration platforms as well as one or more production platforms per hydrocarbon reservoir will be needed in deeper Arctic waters. Fixed structures and floating platforms, with necessary support and rescue systems, will provide the basic alternatives for a water depth range of 50-200 m.

Ice conditions considered in this presentation are moving or landfast one-year old ice, first-year pressure ridges, multi-year floes with ridges, rubble fields, icebergs and islands. Accumulation of ice by adhesion and atmospheric bonding is also discussed. The potential use of fixed offshore structures strictly depends on actual ice loads and effects together with wind, wave and current forces.

In the case of first-year ice with small- and medium-size ridges, modifications of some platforms planned for the North Sea (e.g., TTP and Condeep T-300) could be used. In addition to ice load design, these structures could possibly be equipped with active icebreaking systems. Monopod and monocone types of platforms would resist both first-year and multi-year ice floes having ridges of considerable thickness. Alternative solutions could be based on large gravity structures or platforms having very massive base sections but more easily movable upper parts. Structures of this kind could possibly be used in the presence of rubble fields, large ridges and icebergs. The properties of these ice formations are not known well enough for structural design or possible active icebreaking activity around a structure.

Cost-competitive structural solutions for Arctic water depths more than 50 m require movability and flexibility in depth deployment. Simultaneous use of different building materials, such as concrete, lightweight concrete and steel, as well as hybrid material solutions give new possibilities for the development of Arctic offshore platforms.

ARCTIC OFFSHORE DRILLING AND PRODUCTION SYSTEMS

R.L. Geer
Shell Oil Company

Artic Oil and Gas Workshop

One of the greatest engineering challenges to be faced by the offshore petroleum industry is in exploring and developing the Alaskan Arctic offshore continental shelf where a significant portion of total United States undiscovered oil and gas is expected to exist. The extensive detailed engineering problems for drilling, producing, and transporting oil and gas from such Arctic waters will require the best technologies if the resource potential is to be fully realized and economically produced.

Industry has been actively working in the Arctic waters of Alaska and Canada for a number of years. Alaskan activity, which had its start in Cook Inlet in the 1960s, was primarily in the Beaufort Sea in the 1970s. A number of milestones have occurred in exploration and production in Arctic waters. In 1964, the first of many ice-resistant drilling and production platforms was installed in 90 feet of water in Cook Inlet, Alaska. In 1973, the first exploration well in the Canadian Beaufort Sea was drilled from an artificial island. More advances followed rapidly with the drilling of an exploratory well from a floating-ice platform in 1974 and the first Arctic subsea completion in 1978, also drilled from an ice platform. A new type of artificial island—a caisson-retained island—was installed in 1981.

During the 1980s, Arctic exploration and development activity will increase as a result of the number of state and federal offshore lease sales being held. The various drilling and production systems, which may be used in the 1980s and 1990s in exploring and developing commercial hydrocarbon resources of the major geological basins, will be discussed.

The Long-Term Role of Submarine Transport in The Arctic

L. R. Jacobsen
General Dymanics

Transportation Workshop

Perceiving that by the year 2000, the size of oil and gas finds, supply and demand, and Arctic costs will not deter development in the north, a future role for submarine tankers is predicted. Transport of hydrocarbon liquids by submarines is technically credible, is not technically limited in the Arctic, and has the potential to be the least disruptive to the environment of any transport system in the Arctic.

Modern cargo submarine concepts, dating from 1954, were given impetus by the development of marine nuclear power plants with full recognition of the potential on specific routes, in particular, under ice.

Although a future role for the submarine tanker is not foreseen in the Chukchi and Beaufort Seas, (for several reasons, not the least of which is the shoal water in the predictable offshore producing areas and the proximity of the existing Alaskan pipeline for accepting production from the near shore, offshore areas) there is a basis to predict that, by the year 2010, submarine transport will have a prime role as a link in inter-island transfer of oil and gas among the Canadian island groups. Four scenarios project the submarine performing shuttle tanker service over relatively short routes. For this service, submarines of less than 40,000 DWT capacity will emerge. The submarine will require features to interface with floating terminals. In the event this role materializes, it will foster long lead development of concepts for floating multifunction terminals, cargo loading at the well head, and new marine power plants such as the Marine Fuel Cell and the Marine Gas Cooled Reactor. In summary, the role of the submarine tanker will be as an under-ice transport link integrated with systems of pipelines and/or surface tankers.

Logistics in the Arctic

M. Oshima
Mitsui Engineering & Shipbuliding Co., Ltd.

Transportation Workshop

Brief presentation on future concepts of huge Archimedean Screw Tractor (AST) and Class 7 ice breakers of Archimedean Screw Vessel (ASV) are made below.

Large AST. AST is an amphibious tractor of a new type to be operated in sea ice for transportation of personnel and materials, and for various works such as ice-breaking, oil deployment, etc. AST can break through ice of below designed thickness, and climb up from in-water onto ice when ice is thicker than designed and continue running. An AST of the 10-ton class (AST-002) has been built and has proven its superior performance. A 100-ton, 2,000 HP AST, about 20 meters long and 10 meters wide, will break 3 feet of ice, carry more than 10 tons onboard cargo and tow sleigh-barge/hover-barge.

Large ASV. ASV is a highly efficient ice-breaking vessel suited to shallow water. Archimedean Screws, four in number and attached at the vessel's bow around the water line, interact with ice by means of rotating blades of screws, pulling the vessel's bow onto the ice, thus breaking the ice by the vessel's weight. An ASV of 120 displacement tons, 24.9m long and 7.6m wide, has been built and has proven its superior performance. A 10,000 displacement tonnage ASV, about 135m long and 23m wide, will break 7 feet of ice continuously at 3 knots with required power less than 25,000 HP. It can break through any pressure ridges and clear ice rubbles by ramming operations and get access to the offshore structures surrounded by rubble ice without damaging them. In the 21st century, ASV would be operated in sea ice for supply logistics, pipe-laying, dredging, foundation soil arrangement, anchor handling, oil spill recovery and so on, as one type of efficient, local, ice-breaking vessel.

ARCTIC MARINE ECOLOGICAL WORK

Vera Alexander
University of Alaska-Fairbanks

Scientific Research Workshop

An important trend in the evolution of marine ecological research involves the expanded use of highly coordinated interdisciplinary approaches. To some extent, this trend began for Alaskan waters in conjunction with the environmental information needs of the oil industry. One such interdisciplinary example is the Outer Continental Shelf Environmental Assessment Program. However, the coordination and management between disciplines are relatively weak. Starting with the International Decade of Ocean Exploration in the 1970s, marine scientists increasingly have recognized the efficacy of large research programs in addressing marine ecological problems. In subarctic and arctic waters, with major fisheries and subsistence considerations potentially in conflict with mineral extraction needs, ecosystem research can be of immense importance and great assistance in planning and management. The information generated by programs such as PROBES (Processes and Resources of the Bering Sea Shelf) in the S.E. Bering Sea and PRO MARE (Norwegian Research Program for Marine Arctic Ecology) in the Barents Sea will prove invaluable in the development of Arctic regions and as models for future research.

Spatial heterogeneity and interannual variability in oceanographic conditions are other aspects of high latitude research emphasis, consequently placing Arctic work into a global perspective. Examples include the great influence on biological processes of year-to-year variations as well as long-term trends in the maximum southerly extent of the seasonal sea ice cover. New approaches involving remote sensing and *in situ* instrumentation will assist progress in addressing the impact of such large-scale phenomena on local biological systems and their internal processes.

NON-PETROLEUM ARCTIC MINERAL RESOURCES AND ENGINEERING DEVELOPMENT FOR THE 21ST CENTURY

George Gryc
U.S. Geological Survey

Minerals Workshop

The search for oil (first whale oil and now rock oil) has been the major stimulus for commercial exploration, exploitation and engineering development in the U.S. Arctic. The exploration of Naval Petroleum Reserve No.4 (1944 to 1953) established the feasibility and practicality of a major petroleum and engineering program in the Arctic. Research studies supported by the Naval Arctic Research Laboratory at Point Barrow (1949 to 1981) provided a knowledge base essential to successful living and working in the Arctic. The discovery and development of North America's largest oil field at Prudhoe Bay resulted in still another giant step in engineering innovation.

Minerals, other than oil and gas, have been sought and produced in small quantities in Arctic Alaska. To date, however, that impact on engineering has been minimal. For many years the region north of the Arctic Circle was considered of little or no mineral value apart from the presence of hydrocarbons. However, continued exploration by the mineral industry and systematic geologic mapping by the U.S. Geological Survey and the Bureau of Mines have resulted in one of the most exciting new mineral discoveries in Alaska in the last 25 years. A belt of base metal occurrences, extending over more than 160 km (100 mi) in the Brooks Range, has been documented and ore reserves have been blocked out by drilling. Plans are underway to develop and produce the Red Dog deposit by Cominco Alaska, Inc., and partner NANA Regional Corp., a native corporation established under the Native Claims Settlement Act. Seventy-seven million Mg (85 million tons) of ore have been identified by diamond drilling with grades of 17.1 percent zinc, 5 percent lead, and 2.4 ounces per ton of silver. The engineering systems and structures required for the Red Dog project include an open-pit mine with adjacent milling facilities, a haul road to the coast, 90 to 117 km (56 to 73 mi) long, and a port facility with a causeway and offshore loading platform.

Published estimates of coal resources in northern Alaska—one of the major coal provinces in the world—range from just over 100 billion to 3.3 trillion short tons. The coal rank ranges from lignite A to high volatile A bituminous. Sulfur content ranges from 0.10 percent to 2.0 percent, comparable to that of other western United States coals of similar rank. As the world supplies of oil peak and diminish, it seems inevitable that these Arctic coals will be produced. Studies by the Batelle Institute nearly 30 years ago and recent studies by Bechtel Inc., indicate the feasibility of a marine terminal on the northwest coast of Alaska either for coal or oil or both. Pipelines carrying coal, oil, or a coal and oil slurry have been suggested.

Other onshore mineral commodities of possible long-range commercial development in Arctic Alaska include gold and other precious metals; limestone, including phosphatic limestone; sand and gravel, essential to engineered structures; and black shales with anomalous contents of metals. Offshore mineral potential, north of the Arctic Circle, includes coal, sand, and gravel in addition to oil and gas. South of the Arctic Circle, in the subarctic and bordering the Bering Sea, a host of critical minerals are known to occur but to date, production has largely been limited to precious metals. However, as the engineered infrastructure (offshore and onshore) is developed for the major minerals in the Arctic, development of smaller deposits will follow.

By the year 2000 it is highly probable that, in addition to oil and gas from Arctic Alaska, coal will be mined and shipped to Circum-Pacific nations. Precious metals will continue to be mined. Lead, zinc, and other base metals will be in steady production. All will require special engineering adaptations to meet Arctic conditions, special environmental concerns, and the cultural requirements of the indigenous population.

CONCRETE FOR ARCTIC CONSTRUCTION

A. E. Fiorato
Construction Technology Laboratories

Materials Workshop

Primary objectives of this Symposium are to look to the future, to identify research and development needs and to hypothesize about potential technological advances regarding Arctic engineering. From the viewpoint of concrete materials, a framework for discussion can be centered on the following topics:

1. Current Experience with Concrete Structures
   - harbor structures
   - lighthouses
   - offshore platforms
2. Concrete Materials
   - normal/lightweight concretes
   - high strength concretes
   - durability
   - ice/concrete interactions
   - engineering properties (cold temperature effects)
   - construction considerations
3. Research Needs/Future Developments
   - concrete coatings
   - materials for cold weather repair
   - polymer concretes
   - fiber reinforced concretes
   - composite steel/concrete systems
4. Codes and Standards Developments
   - evaluation criteria for durability
   - design criteria for high strength concretes
   - design criteria for deformation capacity
   - design criteria for composite systems

RELEVANT REFERENCES:

1. ACI Committee 357, "State-of-the-Art Report on Offshore Concrete Structures for the Arctic," American Concrete Institute, Detroit, 1984.

2. "Proceedings of the International Workshop on the Performance of Offshore Concrete Structures in the Arctic Environment," National Bureau of Standards, NBSIR-83-2751, Washington, D.C., July 1983, 67pp.

DEVELOPMENT TRENDS IN STEELS USED IN THE ARCTIC OCEAN FIELDS

K. Itoh
Nippon Steel Corporation

Materials Workshop

This paper describes a variety of properties and future tasks required of steels used for exploration and production systems of oil and gas industries in the Arctic Ocean fields.

The major task is the development of Arctic steels and their software for application (taking the service conditions into account) from the viewpoint of structure safety against brittle fracture, fatigue failure and corrosion problems in the Arctic environment.

In this paper, the trends of Arctic structures, such as drilling rigs, production systems, storage tanks, tankers and icebreakers, and steel pipes for oil-wells and pipelines are summarized. The required properties of Arctic steels used for such structures are noted. Specific steels applies to Arctic structures and trends in the development of steel production technology are described. Finally, future tasks for welding and fracture mechanics, with which the service conditions of Arctic structural steel are determined, are mentioned.

Appendix D:

Technical Presentations

While the workshops were specifically designed to be an informal forum, several participants opted to provide a more detailed version of their remarks by submitting technical papers for publication in the conference record.

Arctic Fixed Platforms for Water Depths Greater Than 50 Meters

K.R. Croasdale
K. R. Croasdale & Associates, Ltd.

Arctic Oil & Gas Workshop (A)

The main reason why offshore platforms are more difficult to design and install in the Arctic than in other regions is because of ice. Ice creates two fundamental problems: large lateral loads, and a restricted open-water season for construction.

In this short paper I will discuss the factors influencing ice forces and how these factors may change in going into deeper water. I will also refer briefly to the experiences to date with fixed platforms in the Beaufort Sea, and what we have learned. This experience plus other current research will be discussed in the context of issues relating to structures beyond the 50 m water depth. Possible concepts for deeper water will be presented, and some comments will be made concerning future research needs.

THE NATURE OF ICE FORCES

When ice moves against a structure it exerts a force. This force is conceptually quite simple to specify; it is governed either by the force required to cause local failure of the ice in front of the structure, or by the environmental driving forces on the ice. The main difficulties in accurately specifying ice loads are to accurately predict the ice failure stress for a given ice thickness; to predict the worst ice thickness to be experienced by the structure; and sometimes to be able to predict the driving forces (if this limits the force).

In this talk, we shall be interested in speculating on how our ability to predict ice loads has and will improve, and on how those parameters governing ice forces may change in going from shallow to deeper water. Before doing so, let's look at ice forces in more detail, because this will help define better the ice characteristics of importance.

LIMIT-STRESS

We can categorize ice loads in terms of three "limit-states." The first I have called limit-stress. In this case it is implicitly assumed that there is an infinite driving force and the ice load is governed by local failure of the ice across the full width of the structure. It can be appreciated that in this case, the ice load is a function of the local ice thickness, the structure width and the effective ice stress or pressure at which it fails and clears. (See Figure 1.)

The ice stress or pressure is governed not only by the ice strength, but also by the shape of the structure which govern the ice failure modes:

- It is, of course, well known that ice will fail in crushing against a vertical structure, and in bending against a sloping structure.

Figure 1. Limit Stress.

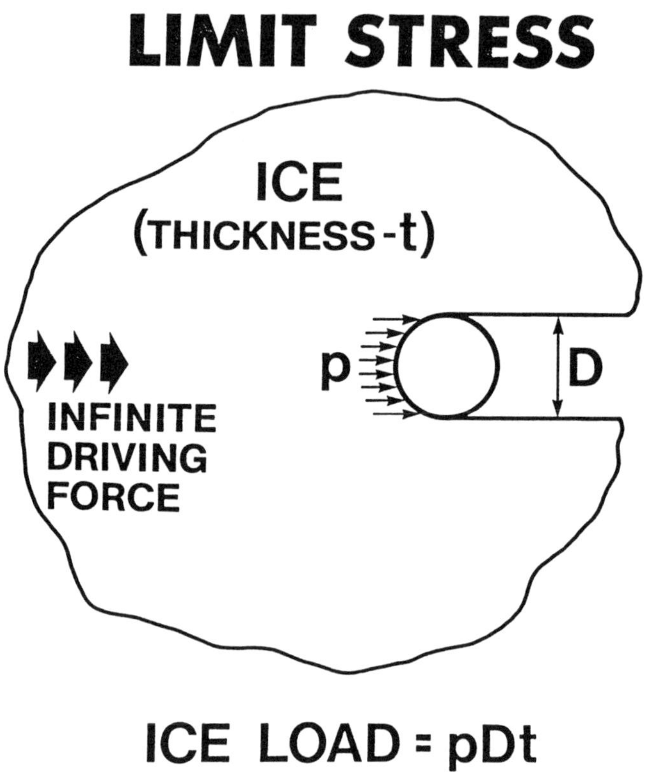

- It is also important, however, to recognize, that broken ice just doesn't disappear but either clears around a structure, or builds up in front of it.
- If the structure is wide and sits on a berm then ice rubble can become grounded in front of the structure. Ice failure is now against the grounded rubble and the shape of the structure becomes irrelevent.

The limit-stress load, accounting for the appropriate ice failure mode, is the usual ice load calculation. The ice load can never be greater than the limit-stress load, but it can be less. (See Figure 2.)

LIMIT-MOMENTUM

One limit to ice loads, for example, which could govern for isolated floes, is calculated from the initial floe momentum. If the floe has insufficient momentum or kinetic energy to fully envelop the structure, then the ice force will be less than calculated using the limit-stress approach. In this case, the floe size and initial velocity have a strong influence on the maximum ice load reached. (See Figure 3.)

Figure 2. Typical Ice Failure Modes Against Structures.

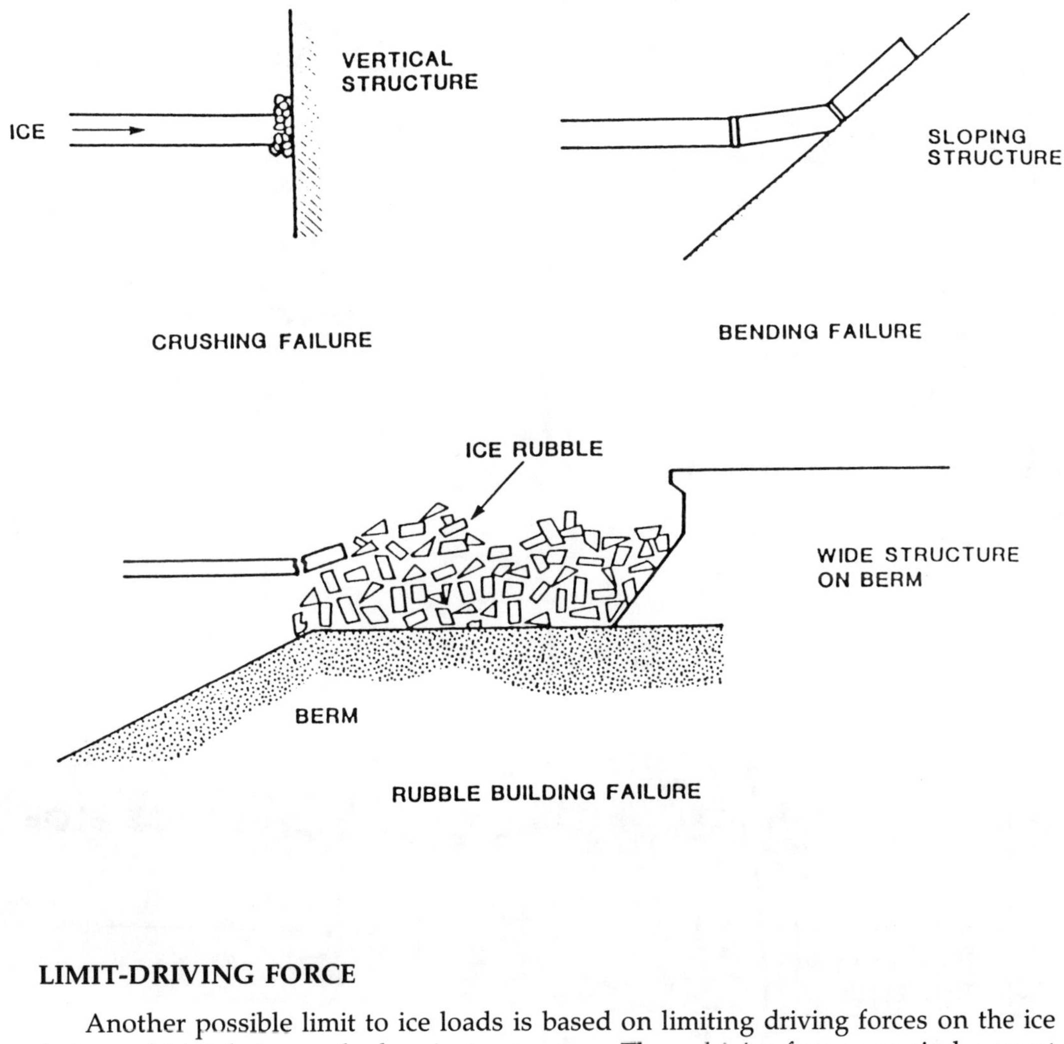

LIMIT-DRIVING FORCE

Another possible limit to ice loads is based on limiting driving forces on the ice feature which is being pushed against a structure. These driving forces are wind, current drag and pack ice forces on the back of the floe. In this case, the floe size and the thickness of the ridging ice have a major influence on the ice forces. (See Figure 4.)

From the foregoing discussion on ice forces, we can now list those ice characteristics which have a major influence on the ice loads. Obviously, ice thickness is of prime importance as is ice strength. Ice strength can be influenced by such parameters as ice velocity, temperature and the volume of ice being deformed. Another important ice characteristic is that the floe size, because this can govern the limit-momentum and limit driving-force ice loads.

How do some of these ice characteristics vary as we go from the shallower to the deeper Arctic offshore regions, in particular in the Beaufort Sea?

Figure 3. Limit Momentum.

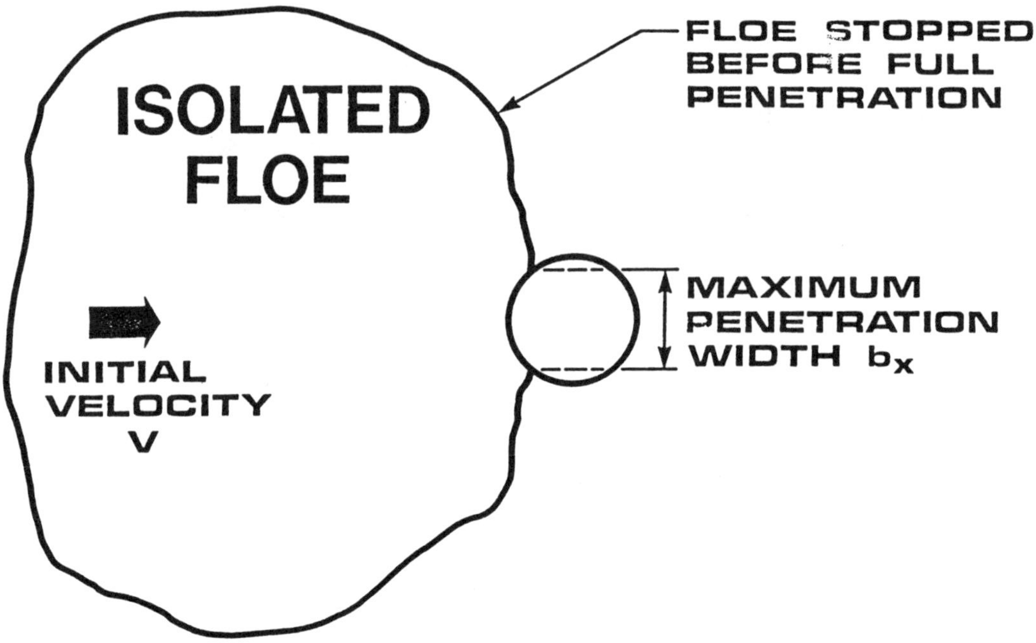

Figure 4. Limit Force.

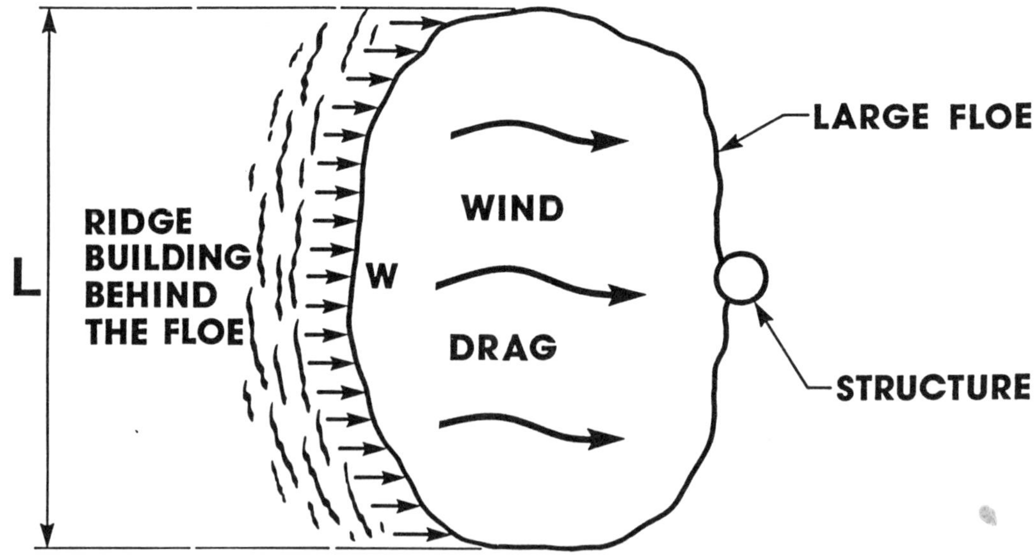

BEAUFORT SEA WINTER ICE CONDITIONS

In less than 20 m of water Beaufort Sea ice will become landfast by January. Beyond the landfast ice and the moving pack-ice is a shear zone, where an open lead often forms under the action of offshore winds. Both first-year and multi-year ridges can occur in the outer edges of the landfast ice, the shear zone and the moving pack. The moving pack will usually be composed of both first-year and multi-year ice. The concentration of multi-year ice will gradually increase with increasing water depth.

Therefore, going from 20 m of water to 100 m, we see no step-change in winter ice conditions. But the probability of encountering thicker ice, and larger floes does increase. (See Figure 5.)

Figure 5. Typical Nearshore Winter Ice Conditions: Beaufort Sea

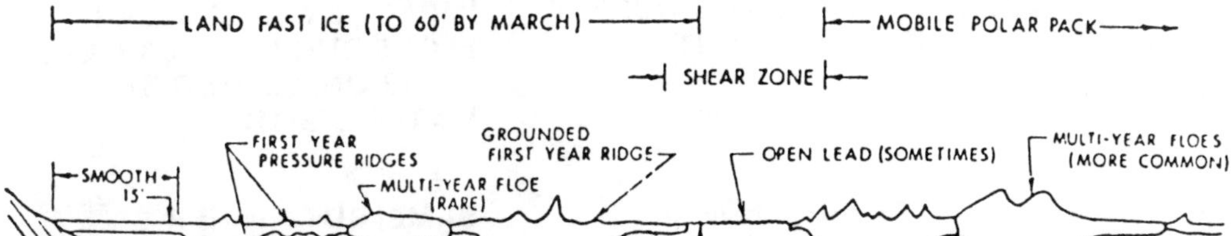

BEAUFORT SEA SUMMER ICE CONDITIONS

In a "good" summer, the nearshore areas of the Beaufort Sea will be ice free. The polar pack will be well to the north of the 50 m water depth. However, in a "bad" summer, the polar pack will be driven south. In fact, the edge of the polar ice can be within the 50 m water depth, threatening even those structures in 20 m of water or less. The edge of the polar pack will normally consist of isolated second-year or multi-year floes, that can be moving up to about 2 knots under the action of winds and currents.

ICE CONDITIONS AND ICE LOADS

We have now set the scene for a discussion of how ice load criteria on Arctic platforms may change in going from less than to greater than 50 m of water. If we look at areas beyond the normal landfast limit of 25 meters, then the actual governing ice interaction scenarios don't change with water depth. They are:

- impact by fast moving multi-year floes in the summer or fall during a polar ice invasion. Such invasions know no water depth border, except when certain extreme thicknesses ground and gouge the sea floor, and are then prevented from further invasion south.
- in the winter, a large multi-year floe, locally thick, can lodge in front of a structure, concentrating the local pack-ice forces on to it. Again, this kind of situation can occur in almost any water depth beyond about 20 m.

So the ice loading cases don't change. What does change, however, is the likelihood of interaction with a certain floe size, thickness and velocity. This likelihood obviously increases with water depth. (See Figure 6.) But it is important to recognize that there is no step-change in ice loading. However, the 100-year ice load, for example, may be higher by 20 to 30 percent in 75 m of water, compared with 40 m (for a given structure).

Figure 6. Arctic Fixed Platforms Ice Interaction Conditions for Design.

ARCTIC FIXED PLATFORMS
ICE INTERACTION CONDITIONS FOR DESIGN

WATER DEPTH	WINTER	SUMMER/FALL
• LESS THAN 25m	RARE RAPID LANDFAST ICE MOVEMENT	IMPACT BY MULTI-YEAR FLOES DURING POLAR ICE INVASION. LIMITED BY WATER DEPTH.
• 25 - 50m	PACK-ICE FORCES CONCENTRATED ON STRUCTURE BY LARGE THICK MULTI-YEAR FLOE LODGED AGAINST IT	SAME, BUT HIGHER PROBABILITY OF LARGER THICKER FLOES
• 50 - 100m	SAME, BUT HIGHER PROBABILITY OF LARGER THICKER FLOES.	SAME, BUT HIGHER PROBABILITY OF LARGER THICKER FLOES AND FLOE COMBINATIONS

So with this point recognized, how can experience with platforms to date in the Beaufort Sea help us with the design of future deeper water platforms? During the past decade, we have seen tremendous progress in the designing and building of structures for the Arctic offshore. From virtually no experience 10 years ago, over 25 locations have been drilled using fixed platforms out to a water depth of 31 m. New designs to withstand high lateral ice forces and short open water construction seasons have been developed and proven. Although ice forces have tended to dominate design considerations, the experience to date with the ice has been extremely favorable. This bodes well for extrapolation of platform use into deeper water.

Let's dwell on what we have learned about ice forces from our experiences to date, and how this knowledge may affect future platform designs. First of all, we should recognize that in the early 1970s, when some of us were struggling with setting ice load criteria for Arctic structures, there were hardly any precedents. Early ice load predictions did not have the benefit of any real-world calibrations. And so, quite sensibly, engineers and scientists developed predictions for ice loads using accepted theories for material deformation and failure applied to ice. The material properties used in these models had to be based initially on small-scale ice deformation tests. Also, the models generally assumed that ice behaved as a continuum, which is the usual and easiest approach in any solid mechanics analysis; and which seemed to be appropriate for the relatively slow moving ice of the near shore.

Considerable research was done by industry in developing these approaches and the early predictions had no benefit from large-scale observations and measurements. It turns out that the early predictions, based on small-scale ice properties and assuming ice deforms as a continuum, were probably quite conservative. Observations and measurements at actual island and structure locations indicate that ice behaves differently than originally thought, and that ice forces are lower than predicted. This is now considered to be due to the fact that:

- ice behaves as a brittle material with a low fracture toughness. The presence of natural flaws in ice may lead to lower average ice strengths than obtained in the laboratory on small samples.
- extreme ice forces need the combination of many ''worst'' events. In a statistical sense the likelihood of all the worst factors occurring together is extremely low. Certainly, to date in the Beaufort Sea, structures have not experienced the worst combination of events.

This last point about the combination of worst parameters is important because it should be emphasized that structures used to date in the Beaufort Sea have not experienced the kinds of ice loading events which might have to be considered in the design of production structures with 20- to 30-year lives. In fact most ice loading events to date on Arctic platforms have been due to first-year ice. As explained earlier, the extreme design condition for Arctic structures is associated with multi-year ice floes perhaps containing ridges. Other extreme, but rare ice features, are hummock fields, and ice islands.

It was recognized some years ago that normal annual ice conditions in the Beaufort Sea do not give us insight into ice interactions likely to govern design. Especially the case of impacts by large multi-year floes in the summer and fall. What was needed was a study site where such conditions occurred annually. Canadian researchers found an ideal study site at Hans Island, a small rocky island in the Kennedy Channel between Ellesmere Island and Greenland. Every year when the ice breaks up, large multi-year floes move down the channel and impact the island. By observing deformation modes and measuring the deceleration of the floes on impact, a great deal can be deduced about the ice forces that a multi-year floe impacting a fixed platform in the Arctic Ocean can exert. The results of these studies are still confidential. But I think it can be said that the ice forces experienced are again much lower than would have been predicted. It is data such as this that will enable future Arctic structures to be designed with increased confidence as well as designed for lower ice loads than previously supposed.

As already discussed, many ice conditions characteristics go into an ice force prediction. For example thickness, velocity, floe sizes, etc. If worst values for all these parameters are combined deterministically then a very conservative ice load will result. But if distribution statistics for these parameters are included into a probabilistic model, the realistic ice forces for various locations can be predicted. Application of this approach is going to be vital as we go from the shallower to the deeper water areas of the Arctic. (See Figure 7.)

Figure 7. Incorporation of Ice Conditions and Large-Scale Calibrations into Statistical Approach to Ice Loads.

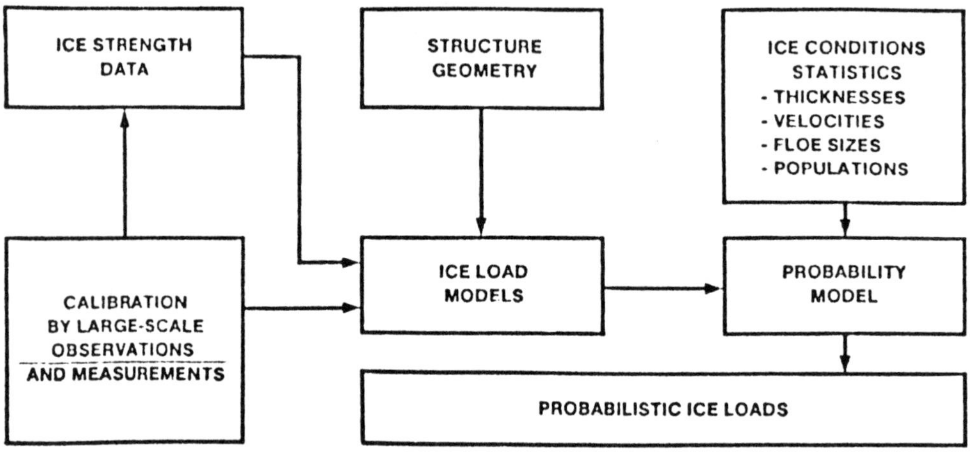

ICE ISLANDS

A word about very rare but extreme ice features such as ice islands. Ice islands are pieces of ice shelves which periodically calve off the coast of Ellesmere Island. When they calve they can be 50-60 m thick and tens of km across. These ice island fragments, grounded off the Alaskan coast, were about 1 mile across and 30 m thick. The chance of a very large ice island, say 10 km across, moving into the coastal areas is considered very unlikely.

In fact simple collision models can be developed based on future ice island calvings. These probability models indicate that the chance of collision is very slight. This, together with the recognition that ice islands can be detected months ahead and marked, suggests that ice islands probably shouldn't be designed. Early detection and tracking would allow oil wells to shut-in and people to evacuate from the platform well ahead of any collision. With respect to global ice loads then we can summarize the following:

- there is no step change in governing ice interaction conditions as we go further offshore
- the two governing cases are impact by multi-year floes in the open water period and pack ice forces concentrated on to the structure by a large floe in front of it during the winter
- the probability of larger and thicker floes increases with water depth
- the overturning movement increases with water depth
- experience to date and current research suggest a downward trend in ice forces for a given condition.

CONSTRUCTION SEASON AND SEA FLOOR CONDITIONS

So far I have covered the issue of ice conditions and ice loads. The other two important issues are construction season and sea floor conditions.

There is not much to say about construction season except that the open-water season gets shorter as one goes further offshore. But variations from year to year are probably as great as variations from location to location. Nearshore in the Beaufrot Sea the typical open-water season is about 70 to 100 days. To date, most open-water operations, such as dredging, have had to be conducted within that period. Although some operations, such as floating drilling, have been extended into freeze-up, using ice-strengthened vessels and ice management techniques. The same kinds of season extension could also be envisaged for dredging and construction.

But my prediction is that fixed platform design will tend to favor those concepts that can be installed within the short open-water season. Further offshore, such operations might be designed to occur in light ice conditions. This philosophy seems to have been accepted by most operations and designers to date who have moved increasingly towards designs for rapid deployment with minimum foundation preparation.

I would like to say a few words about sea floor issues, even though this topic has been well covered in the articles in the special issue of the Marine Technology Society *Journal* by Brian Watt and Bob Rea. To summarize:

- the shelf break in the Beaufort Sea starts at about 65 to 80 m of water
- the thickness of weak holocene sediments increases with water depth
- massive slide features have been observed beyond about 65 to 80 m
- shallow gas and decomposing hydrates may be a problem
- there is earthquake activity in the Eastern Beaufort Sea areas.

PLATFORM CONCEPTS

Finally, what about some of the concepts for fixed platforms that might be used in the deeper waters in the Arctic? Some of the potential concepts include:

- caisson berm systems
- monopods
- stepped caisson structures
- caissons with sloping sides
- cone structures, both upward breaking and downward breaking

Personally I don't have any new ideas or favorites. Furthermore, I think all the geometrical alternatives (these and others) have probably already been suggested and publicized at some time during the past decade. Many of these alternative concepts are described in the articles on arctic engineering, by Michael Rojansky and John Bruce in the 1984 First Quarter issue of the Marine Technology Society *Journal*. I don't want to repeat what they have said about them, but let me say a few words about each, especially with respect to ice loads.

CAISSON-BERM SYSTEMS

Simple caisson structures sitting on, or surrounded by sand or gravel berms, are a natural extrapolation of some of the existing technology being used at present in the

Beaufort Sea for exploration. In areas where sand or gravel are locally available, such a concept has the following advantages:

- experience with similar systems currently in use can be exploited in their design
- ice rubble which builds up around them can provide a cushioning for severe ice impact conditions and mitigate the forces
- even without rubble fields, the concept can be configured to have good energy absorbing characteristics, especially against thick ice features which strike the berm first
- with a vertical-sided structure there should be no ice ride-up problems and the area to be reinforced against local ice pressures is minimized
- the sand berm increases the weight on the sea floor and will, over time, improve the soil strengths.

On the other hand, there are several disadvantages including:

- the large dredging operations requiring several seasons for construction
- wave erosion problems of the berm mean that annual maintenance may be needed, but this will depend on the depth of the berm below sea level
- the large sand mass on the sea floor may lead to excessive settlement
- the extensive rubble field may restrict access during the winter months to supply ships and tankers
- the berm may be susceptible to instabilities during earthquakes.

MONOPOD

At the other end of the spectrum of concepts is the monopod that attempts to minimize the ice forces by keeping the structure as narrow as possible. This is the major advantage of the concept. Others include:

- small perimeter at the ice-line to reinforce against local ice pressures
- no ice ride-up and clearing forces.

Disadvantages of the monopod for permanent production structures include:

- ice loads, due to thick ice features, are currently not easily specified. Ice pressures are likely to be high because the geometry of the crushing zone leads to a high degree of confinement
- structurally, the concept is difficult to design for, depending on the shaft diameter and water depth
- dynamic excitation of the shaft/deck structure can be expected under fluctuating ice forces. This can lead to dynamic magnification of ice forces and possible fatigue problems
- the overturning movement can be higher than with other concepts.

STEPPED STRUCTURES

A variation on the vertical-sided monopod is the stepped structure, recently suggested by Ben Gerwick. This arrangement is perhaps structurally easier to implement and should not be subject to dynamic excitations. It also has a better ability to absorb impacts due to thick ice fractures, because of the gradual pick-up in the ice force as the various steps of the structure are contacted by the ice feature. Overturning movements with this structure may also be less for thick ice features because of the tendency for the thick ice to contact the lower steps first.

SLOPING CAISSON STRUCTURES

It has long been known that a sloping structure can fail the ice in bending at lower loads than crushing. This is certainly true for a narrow structure against which crushing pressures can be very high. Recent large-scale data, however, such as from Hans Island, suggest that ice failure across a wide structure even when vertical-faced may be lower than supposed, and this new information may undermine the advantages of a wide sloping structure.

However, wide sloping structures do offer the potential for lower initial ice loads when calculated using conventional wisdom. But care is needed with these structures for the following reasons:

- the sloping geometry causes ice ride-up which must be allowed for in the design of the upper parts of the structure
- if the ice does not clear, but falls back on the advancing ice, then ice loads are likely to be higher than calculated, assuming simple bending failure
- if ice freezes to the slope of the structure during stationary periods, then ice loads may be higher than calculated from bending failures. Special coatings, which may be difficult to maintain, may be needed to mitigate this problem.

CONICAL STRUCTURES

The problem of ice ride-up against the superstructure and deck can be avoided by making the structure into a cone and raising the deck. The cone shape, with its narrow neck, is much better in terms of ice clearing than the wide sloping structure. All upward breaking sloping structures do have the potential advantage of a downward component of ice force which reduces the overturning movement and can increase sliding resistance. But, of course, this advantage only prevails if the bending mode is maintained. The previously mentioned problems with adfreeze still apply.

A downward breaking cone may have some advantages, for example:

- clearing forces may be lower
- adfreeze forces may be reduced.

But several potential disadvantages exist, namely:

- ice may jam in the neck region
- the overturning movement will be higher than a conventional cone
- structurally, the concept may be more difficult to design and, therefore, may be more costly
- depending on the diameter of the neck, dynamic problems may exist.

CONCLUDING REMARKS

To conclude, I would like to emphasize the following points:

- there is no shortage of imaginative concepts for arctic fixed platforms

- structural optimization once the environmental loading and foundation conditions are known, is fairly straightforward
- the challenge in assessing feasibility and optimizing these concepts is still that of better quantifying ice loads and foundation conditions
- ice loads, based on small-scale ice properties and continuum models, are probably too conservative
- measurements and observations at existing Beaufort Sea exploration structures are helping to cultivate ice load predictions
- but normal ice conditions in the near-shore Arctic do not represent extreme design conditions, especially for production structures and those in deeper waters
- nevertheless, we expect that as more experience is gained and more research is done, the design ice loads for a given condition will reduce
- there is a need to gather data on real-scale ice interactions which simulate extreme ice loading events on structures
- the work at Hans Island did this for the impact of multi-year floes in the summer against wide vertical structures
- I believe there is a need for similar work on winter pack-ice forces; with particular focus on limiting-driving forces within the pack ice
- enough data on ice statistics need to be gathered to enable probabilistic methods for ice loads to be applied with confidence
- there is no step change in ice loads in going into deeper water, but it generally increases
- however, sea floor problems at the shelf break, which is beyond about 70 m in the Arctic Ocean, may limit the practicality of bottom-founded structures.

In any case, given a recognition of these points, and the will and resources to address them, fixed and floating systems in the deeper Arctic waters can become realities.

REFERENCES

Croasdale, K.R. and R.W. Marcellos. "Ice and Wave Action on Artificial Islands in the Beaufort Sea." *Canadian Journal of Civil Engineering,* Volume 5, No. 1. 1978.

Croasdale, K.R. "Designing Offshore Structures for Ice." Coastal Engineering Professional Development Seminar, Queens University, Kingston, Ontario, Canada. April 12–13, 1984.

VIEWPOINTS OF THE USE OF FIXED PLATFORM STRUCTURES IN DIFFERENT ICE CONDITIONS OF THE ARCTIC

Pauli Jumppanen
Technical Research Centre of Finland

Arctic Oil & Gas Workshop (A)

Since the early 70s more than 20 artificial islands have been successfully built in the American and Canadian Beaufort Sea where water depths are less than 20 m. New construction techniques and slope protection methods have been developed especially in connection with the Issungnak and the Mukluk projects.[1] For the completion of Issungnak Island, large scale hydraulic dredging of the surrounding seabed took place, while in construction of Mukluk Island, both transportation of gravel in the winter to a temporary stock, and barge transportation in the summer to the building site were performed. Latest developments of superdredgers will give a possibility for even more effective construction methods.[10]

The need to reduce the amount of sand and gravel in islands and underwater berms led to the design of caisson-retained islands and caisson-type structures which have been successfully used in the Canadian Beaufort Sea since 1981.[8] The completed caisson-retained islands are the concrete caisson Tarsiut of Dome Petroleum and the Steel Caisson Retained Island (SRI) of Esso. Caisson-type structures built so far are the Single Steel Drilling Caisson (SSDC) of Dome Petroleum, the Mobil Arctic Caisson (MAC) of Gulf and the Concrete Island Drilling System (CIDS) of Global Marine. These structures have been designed for water depths of less than 30 m but dimensioned for heavy ice loads including multi-year ice features of several types.

The first floating drilling vessel suitable for winter drilling operations in difficult ice conditions was Gulf's Canonical Drilling Unit, Kulluk.[9] The Kulluk (structure and mooring system) is designed for ice thickness of 1.2 m and can operate in water depths of up to 55 m. Drilling season can be extended further by using icebreaking support.

In addition to the completed projects mentioned above, several more or less carefully studied offshore structures or structural concepts have been proposed for use in deeper waters, especially in the Beaufort Sea and around Alaska. A short list of eight different concepts of fixed Arctic platforms is presented below:[2,3,5,7,11,12]

- Monopod Jackup Drilling Rig (steel structure, water depth 5-27 m)
- Sohio Arctic Mobile Structure, SAMS (concrete/steel structure, water depth 12-18 m)
- Recoverable Arctic Platform, RAP (steel/concrete structure, water depth up to 30 m)
- Sonat Hybrid Arctic Drilling Structure (steel/concrete hybrid structure, water depth ≈ 20 m)
- DASF Drilling Production Structure (steel structure, water depth up to 30 m)
- Arctic Cone Exploration Structure, ACES, (lightweight concrete/steel structure, water depth 15-33 m)

- Stepped Pyramid Structure (concrete or steel-concrete hybrid structure, water depth 50-200 m)
- Gravity Base Production Platform, GBS/Hibernia (concrete structure of the North Sea type, water depth up to 100 m).

The characteristic features in the structural systems described above are:

- movability based on water ballasting
- flexibility in depth deployment
- use of more concrete and lightweight concrete than in nonarctic offshore structures
- use of steel-concrete combinations and, especially, hybrid material solutions.

ENVIRONMENTAL FACTORS

In the development of Arctic offshore structures for water depths of more than 50 m, basic environmental considerations should be made on:

- ice loads and ice-structure interaction
- icing and spray ice formation phenomena
- wind, wave and current parameters
- soil conditions in the sea bottom.

BASIC ICE FEATURES

Ice conditions in different parts of the Arctic have been discussed in detail in various papers and publications. In the following, ice phenomena have been divided into seven basic types which are described in Table 1. The characterization of ice types is very broad and deals only with further considerations in this paper.

Of the ice features described in Table 1, annual ice formations are well-known in most parts of the Arctic seas. This is mainly caused by icebreaking activities needed on the northern sea routes and sub-arctic seas in wintertime. The ice parameters of multi-year winter and summer floes are also rather well known in some parts of the American and Canadian Arctic, especially in the transition zone on the Beaufort Sea. Most of the multi-year pack ice region is quite poorly known. There is no good characterization either for multi-year ridges in floes or rubble fields around large structures or other obstacles. Iceberg flux and characteristics have been studied extensively in the Labrador Sea, especially in the Hibernia field, but not much iceberg data are available from other parts of the Arctic.

WIND, WAVES AND CURRENTS

Wind, wave and current parameters are generally better known than ice conditions, due to long term collection of meteorologic data. Typical wind speeds on Arctic seas are 40-50 mps (one hour maximum) and 60. . .70 mps in gusts. Wave heights and periods depend naturally on the water depths. Typical maximum value of wave height is 15-20 m with a period of 8-10 sek for water depths between 50 and 100 m. The extreme speeds of surface currents are often 1.0-2.0 mps when the corresponding values of bottom currents are normally 70-80% lower. But one should remember that there are

Table 1. Basic Ice Features in the Arctic.

FIRST-YEAR LANDFAST ICE —ice thickness, max. 2 m —in ridges, typical max. 10-20 m, extreme 30 m —ice movement, max. 5 m/hr	
FIRST-YEAR MOVING ICE —ice thickness, max. 2 m —in ridges, max. 5 m —ice movement, max. 1.5 m/s	
MULTI-YEAR PACK ICE —ice thickness, average 3.5 m, typical max. 10-12 m —in multi-year ridges, typical max. 30 m, extreme 50 m —ice movement, 0.15 m/s	
MULTI-YEAR SUMMER FLOE —ice thickness, typical max. 7-10 m —ice movement, 1.2 m/s	
RUBBLE PILES —ice thickness, typical max. 20-30 m	
ICEBERGS —depth, max. 250-300 m —weight, max. 20-50 milj.tn —movement, 0.5 m/s	ICE ISLANDS —thickness 50-60m —diameter 10-50 km

quite large local variations in all environmental factors, especially in current parameters. Tidal range is quite large (e.g., in the Newfoundland region and the southern coast of Alaska), and should therefore be considered as a design parameter.

GEOTECHNICAL CONDITIONS

The design of fixed offshore structures presupposes detailed knowledge and good characterizations of seabed morophology, soil types and properties and soil-structure interaction behavior. The basic soil types in Arctic seas are silts, silty clays, clays, silty sands and sands. Silts and clays can be more or less consolidated than sands in loose or compact condition. Existence of surfacial bedrock or shallow underlaying permafrost is also possible in certain areas.

Resistance to slippage has been the main problem in the foundation design of shallow water structures. In water depths of more than 50 m, the load-bearing capacity of soils will play a more important role. It seems that special foundation engineering methods, such as the use of skirts, piles or thermal piles, soil replacement, chemical stabilization of soils, etc., are needed in most cases.[4] Improvements are also needed in geotechnical underwater testing. Equipment for remote controlled sampling, data transmitting to the surface and analyzing *in situ* are under development.

DEVELOPMENT OF ARCTIC PLATFORMS

Design Philosophy

Offshore structures used in Arctic conditions must satisfy very high safety requirements. This is due to environmental risks and governmental policies designed to protect the fragile Arctic environment. The selection of structural types and construction technology, however, is made on the basis of total economy considerations in which manufacturing, construction, maintenance and operational costs are taken into account. The main task of the structural design is, therefore, to guarantee the required safety level of both the total structural system and its essential components.

The platform design includes, a.o., the calculation of overturning moments, studies on foundation stability and dimensioning of various structural sections. In addition, attention should also be paid to the durability of building materials and components. The behavior of materials is greatly influenced by the corrosive sea atmosphere, large temperature variations, freezing and melting cycles, and wearing and scouring effects of moving ice. Special protection methods, both underwater and above the surface, can considerably increase the durability and lifetime of a platform structure.

New Technologies

The future development of Arctic platforms will also depend on the progress made in other fields of technology. Important branches of marine technology that should be considered in this connection, include the following:

- weather information and ice forecasting methods with ice load control and warming systems
- ice management techniques, including active and passive icebreaking and clear-up systems in connection with offshore structures
- underwater technology in construction, drilling, production, and repair and control.

Several types of underwater systems will provide the basic alternative to platform structures in water depths beyond 50 m. In the exploration phase, subsea facilities are usable in different ways to complete operational capabilities of an offshore platform. In the production phase, subsea systems can dramatically reduce the need for production structures. Furthermore, it will be possible to build a field based totally on remote controlled underwater production facilities. Development of fully automatic operational systems for drilling, for example, can take 20 years or more. However, most of the facilities which today are under design and testing, will be available and ready for use during the 21st century.

FIXED PLATFORMS FOR DIFFERENT ICE CONDITIONS

Environmental conditions, especially ice loads and soil properties of the sea bottom, provide the basic criteria for the selection of Arctic seas (Figure 1). To discuss structures adapted to different ice conditions, the Arctic region is broadly divided into three different zones called:

- sub-arctic or marginal zone
- multi-year zone
- iceberg zone

The border lines of the sub-arctic zone are formed by maximal extensions of annual and multi-year ice cover. However, the latter boundary is not very well known in most parts of the Arctic. In the sub-arctic zone there are only first-year landfast or moving ice formations (Table 1).

The multi-year zone covers the rest of the Arctic, except the regions where icebergs or ice islands (tabular icebergs) are found. Multi-year ice features are mainly controlled by movements of the Arctic pack and they comprise different types of floes with ridges and rubbles (Table 1). The following considerations about the use of fixed arctic platforms will be made on the basis of the presented ice zone categorization.

Sub-Arctic Zone

The sub-arctic zone includes several Arctic seas, e.g., the Bering, the Barents and the Ohotsk Seas (Figure 1), where oil and gas exploration activities are already taking place. Exploration drillings can be carried out by using conventional rigs or arctic platforms (semi's, jack-ups and drillships) specially designed and equipped for extended drilling seasons. Ice-resistant floating drilling units, such as the Kulluk, would also be adapted for year-round operations.[9]

Figure 1. Boundary Lines of Different Ice Zones in the Arctic.

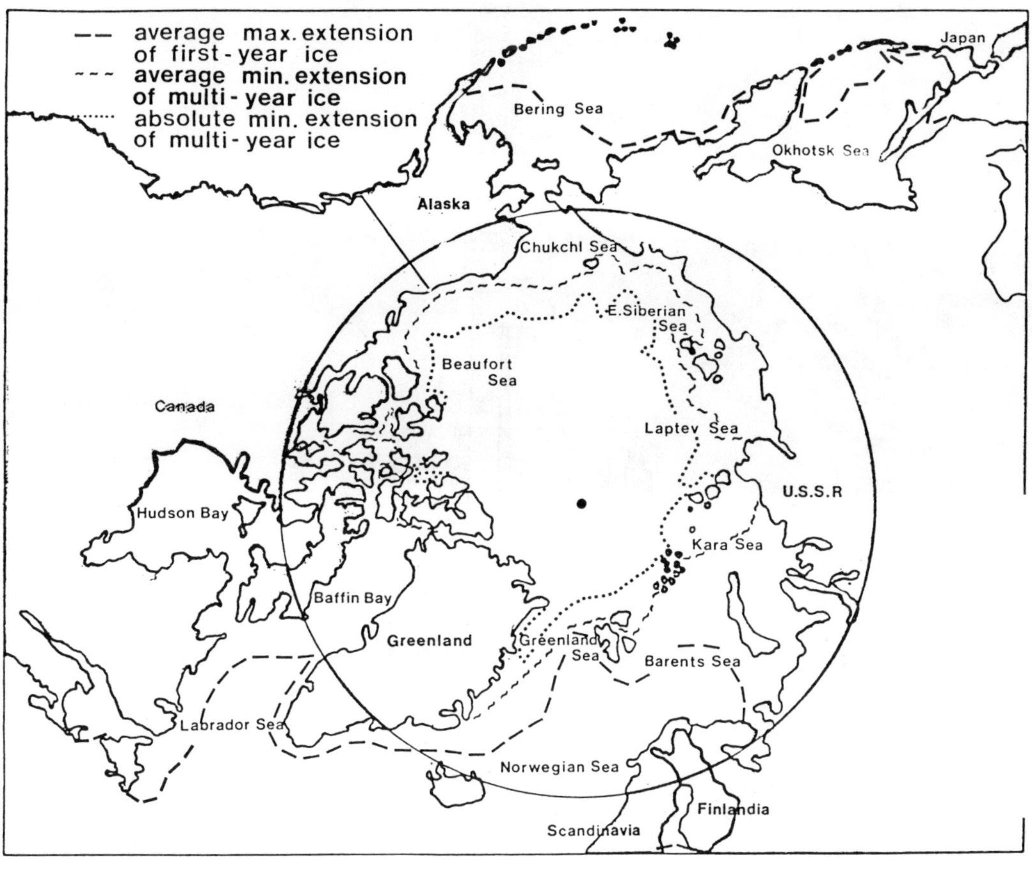

Oil and gas production in sub-arctic seas and in water depths beyond 50 m, could be based on underwater production systems together with fixed or floating maintenance platforms. A very successful development of Condeep concrete platforms made for the North Sea fields has proven the potential possibilities of fixed gravity structures.[6] Wind and wave loads on the North Sea are nearly twice as high as the common loads on the Arctic seas.[13] Thus, these platforms are also capable of resisting some ice loads.

The North Sea types of platforms can then be converted and designed to be used in light or moderate ice conditions of the sub-arctic seas. Figures 2a and 2b present the known Condeep T-300 and Heerema TTP platforms equipped with active icebreaking and ice clear-up systems. Furthermore, Figures 3a and 3b show two possible Arctic modifications of the tripod platform, where ice loads are reduced by a wall framing cone. By the use of special ice load design and effective ice management systems, the platform types presented above could well be applicable in most parts of sub-arctic seas. In the case of very heavy ice cover, gravity structures drafted in Figures 4, 5 and 6 could be selected instead of the North Sea type structures.

Figure 2a. North Sea Platforms—Condeep T-300 Equipped with Air Bubbling System.

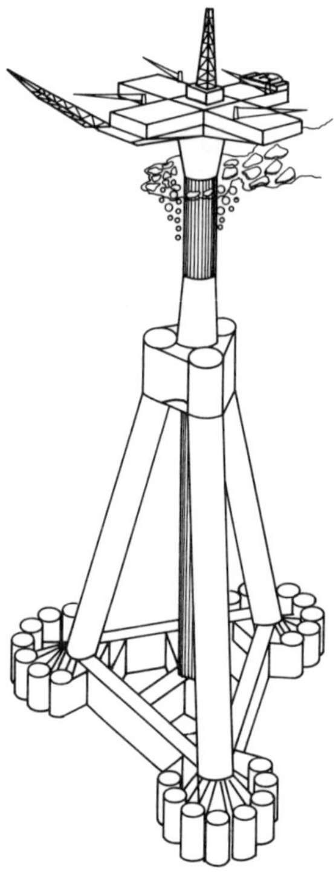

MULTI-YEAR ZONE

In the multi-year ice region, the use of floating platforms is restricted by a short ice-free period and the risk of extensive summer floes. In addition, the handling of floating rigs in winter ice conditions is quite impossible. Nevertheless, ice-resistant

Figure 2b. North Sea Platforms—TTP Equipped with Icebreaking System.

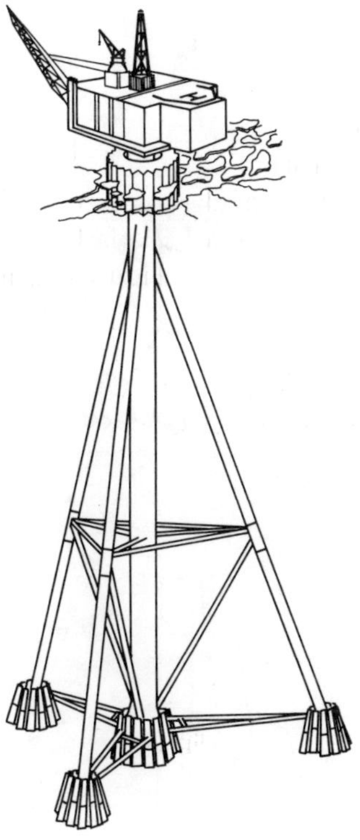

floating platforms are usable for exploratory drilling under the extended summer drilling season. However, most exploration drillings will probably be carried out by using movable gravity platforms. Basic types of this kind of structures are described in Figures 4, 5 and 6.

Because of huge ice loads, possible permanent production platforms in the multi-year zone could be very large, massive gravity structures. The high price of such structures will certainly lead to an increased use of underwater production systems. One possible solution would be the building of only one or two gravity structures per hydrocarbon field and, in addition, extensive utilization of underwater production facilities. A second alternative would be a remote controlled, totally automatically operating subsea production system.

ICEBERG ZONE

Icebergs and ice islands can have dimensions of several kilometers and weight up to hundreds of millions of tons. There are no economically feasible structures which could take ice loads caused by the biggest moving icebergs. On the other hand, icebergs up to a weight of 1-3 mill. tons can be towed, and the towing technique will be further improved during the next decades. All this leads, undoubtedly, to a movable technology in cases of both floating and fixed platforms.

Figure 3a. Arctic Modifications of the Tripod Platform.

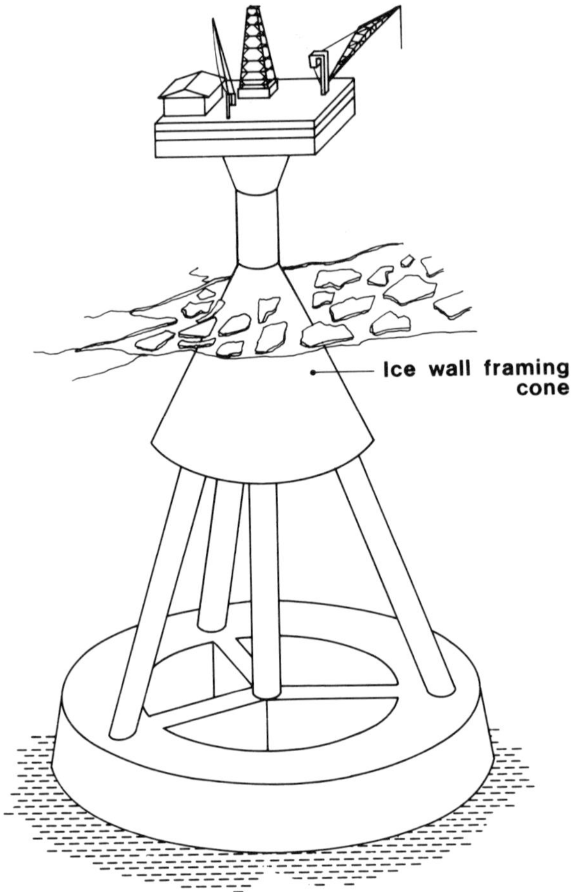

In the iceberg region, exploratory drillings can be carried out almost totally from conventional platforms. The drilling must be discontinued and the platform rapidly moved in the presence of a big iceberg. It is also possible to use floating production platforms in the iceberg zone. In the areas, where the bottom scour of icebergs takes place, subsea systems (e.g., wellheads) need to be protected in a special way.

Fixed platforms adapted to the iceberg zone should also be movable in critical situations. The Condeep-type platforms can easily be designed to satisfy these requirements. Another alternative would be a fixed platform having a gravity base and movable upper part. Concepts of this kind of structural combination are presented in Figures 7 and 8. The movable upper part could be positioned dynamically and could also include some share of the production facilities. The fixed gravity base should be designed for different effects of moving icebergs and could contain most of the production equipment.

Figure 3b. Arctic Modifications of the Tripod Platform.

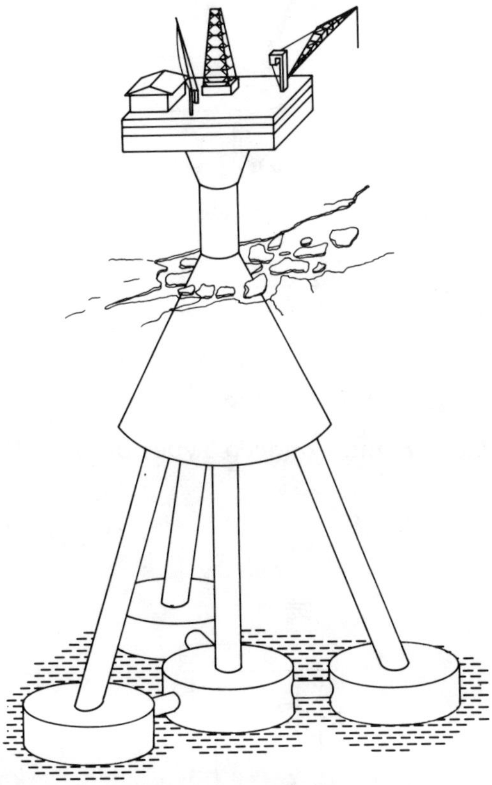

Figure 4. Caisson-Type Gravity Platforms.

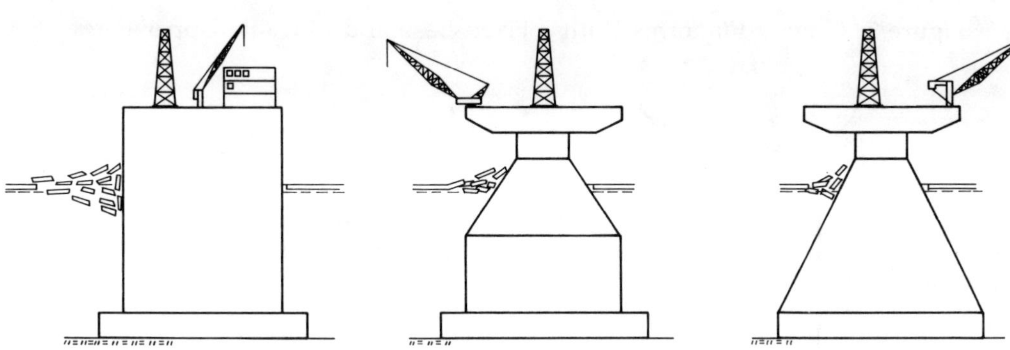

Figure 5. Fixed Platform Types: Stepped, Cone and Double Cone.

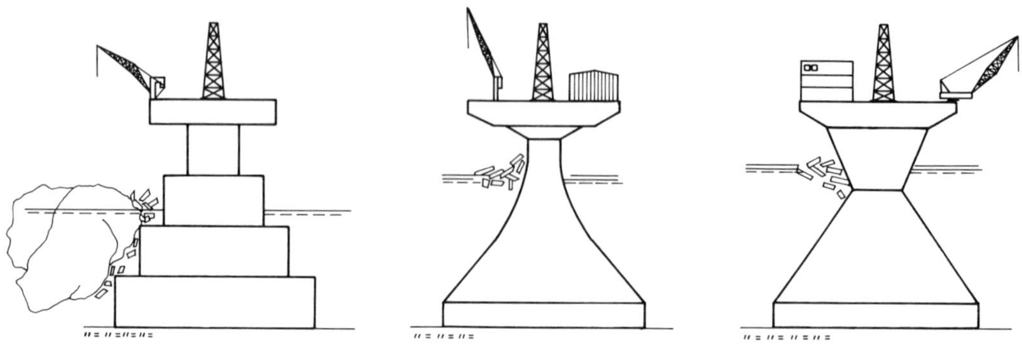

Stepped Cone Double Cone

Figure 6. Jack-up and Condeep Types of Fixed Platforms.

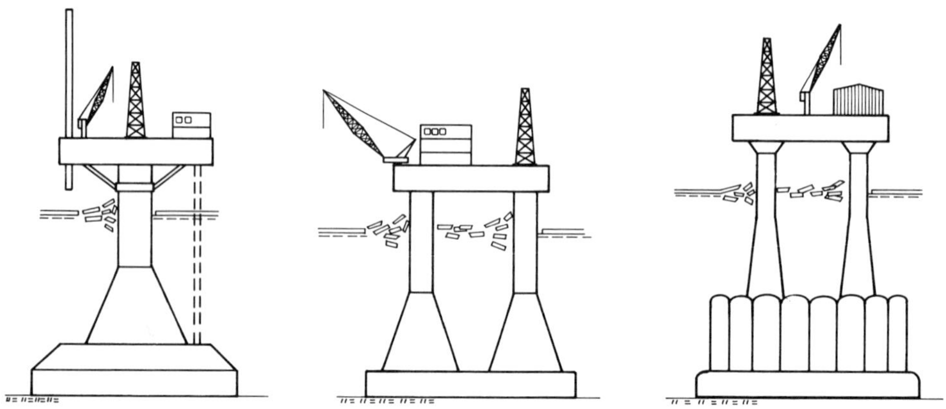

Figure 7. Gravity Platforms With a Fixed Base and Movable Upper Part.

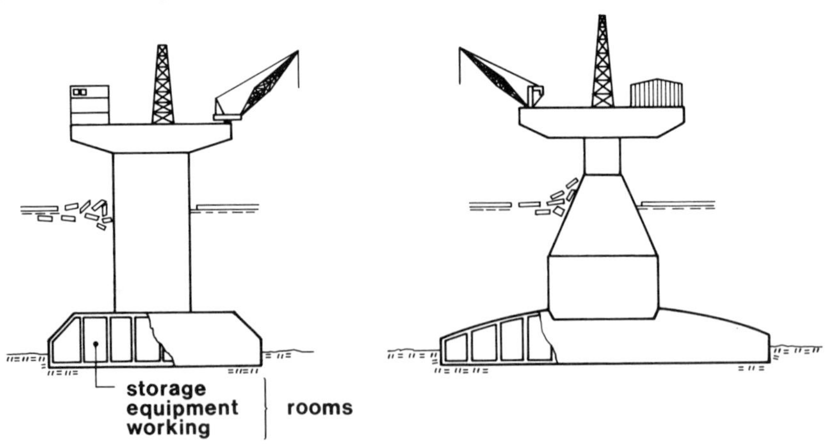

Figure 8. Platform Structures With Large Gravity Bases and Floatable Upper Parts.

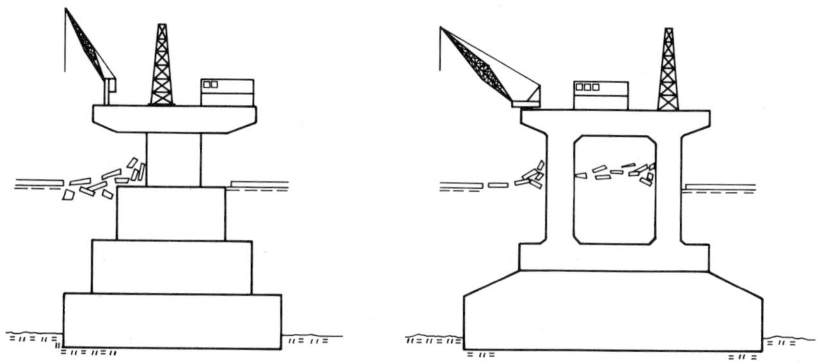

CONCLUSIONS

The previous discussions indicate that several types of fixed platforms can be developed in future to be used in deeper waters for both explorations and production purposes. The North Sea platforms can be modified to survive in light or moderate first-year ice conditions. Caisson, monopod and multipod structures are also adaptable to multi-year ice regions. These kinds of large gravity structures are very expensive and, therefore, only a few platforms per hydrocarbon field can be built. The structures will be optimized together with the use of underwater exploration and production facilities.

Most of the proposed gravity structures have a conical form to permit bending or shearing failure in ice rather than complete crushing. This would reduce the maximal ice loads by as much as 50-70%. However, icing of a structure, ice accretion and ad-freeze would certainly change the expected failure mechanism. Methods to prevent these phenomena will be of great interest in the future development of fixed arctic platforms.

REFERENCES

1. Ashford, R.A. Engineering and Construction of Mukluk Island. 16th Annual OTC (4670), Houston. 1984.
2. Bhula, D.N., J.N. Birdy and F.J. Bruen. Design of Concrete Gravity Structures to Withstand Concentrated Ridge and Flow Impact Loads. 16th Annual OTC (4708), Houston. 1984.
3. Bruce, J.C. and W.D. Roggensack. Second Generation Arctic Platforms. 16th Annual OTC (4798), Houston. 1984.
4. Buslov, V.M., D.R. Williams, R.A. Hayes and H.R. Bivens. Bottom Preparation for Gravity Structures in the Beaufort Sea. 16th Annual OTC (4796), Houston. 1984.
5. Byrd, R.C. et al. The Arctic Cone Exploration Structure. 16th Annual OTC (4800), Houston. 1984.
6. "Fixed Platform Viewed as Best Bet for Troll." *Offshore*, December. 1982.

7. Graca, A. D. New Concept of Offshore Drilling/Production Arctic Platform and Dynamic Analysis of the Fast Removal Operation. Proc. of the Third Int. Offshore Mechanics and Arctic Eng. Symposium, Vol. 3. 1983.
8. Hnatiuk, J. Current and Future Offshore Activities in Canada. 16th Annual OTC (4704), Houston. 1984.
9. Loh, J.K.S., J.C. Stramberg and K.P. Cusack. New Generation Arctic Drilling System. 16th Annual OTC (4797), Houston. 1984.
10. Noble, P.G. Arctic Offshore Engineering—An Overview. Marine Technology, Vol. 20, No. 4. 1983.
11. Padron, D.V., F. Joneidi and J.F. Nixon. A Deepwater Actively Frozen Seabed (DAFS) Drilling/Production Structure for the Beaufort Sea. 16th Annual OTC (4705), Houston. 1984.
12. Rojansky, M. ''Arctic Exploration and Production Structures.'' MTS *Journal*, Vol. 18, No. 1. 1984.
13. Simiu, E. and S.D. Leigh. Turbulent Wind Effects on Tension Leg Platform Surge. NBS Building Science Series 151. 1983.

Required Technology for the Development of Arctic Oil and Gas Reserves

J. Preston Mason
Seaflo Systems, Inc.

Arctic Oil and Gas Workshop (A)

World demand for oil, gas and coal has been given as the driving force for development of the Arctic regions. If that is the case, the timing of Arctic development depends on the worldwide economic climate, trends in energy conservation, the success of research in nuclear fusion and other alternate energy sources and, no doubt, other factors. Rather than address the mind-boggling problems of predicting the interaction of these factors, I will assume here that there is a demand for oil and gas reserves in the Arctic and address the technology required for their development.

Reserves located in water depths of 300 ft., and perhaps even deeper, will probably be developed from fixed or portable bottom-founded structures designed to resist ice forces by some method. This is a straightforward extension of proven ''conventional'' Arctic methods. The problem of development by these methods is primarily transportation, logistics and cost. Beyond those water depths, production operations will be done on the surface from some form of floating facility (including surface ice in some areas), between the ocean surface and the seafloor (probably on the seafloor), or from beneath the seafloor.

Development from a moored floating surface facility requires use of seafloor wells and requires that the surface facility be kept on station within the elasticity of the connection between the surface and the seafloor.

Seafloor well technology, developed for use in deepwater, ice-free areas, is directly applicable to the Arctic—if the logistics and transport of the seafloor equipment, and the well drilling and completion methods can be developed.

Present methods of exploration, drilling of development wells and completing wells on the surface require that the surface station stay within a few percentages of water depth from the well site. In spite of a great deal of research into stationkeeping methods, conventional floating drilling and production facilities can stay on location only when the surface ice does not move for the required operational period, or when the ice is thin enough that a practical mooring or dynamic positioning system can resist the ice forces on the vessel. Since these conditions exist only in a small part of the Arctic, production systems that tolerate greater motion of the surface station will have to be used to develop hydrocarbon reserves in most of the Arctic.

One approach is the use of high-pressure flexible pipe risers that have been developed for ice-free areas. Practical handling methods, to let out or take up riser length to allow the surface facility to drift relative to the wells, will be a major problem. Obviously, the better the riser handling method, the farther the surface facility can be allowed to drift from the neutral position and the less demand there is on the stationkeeping method. Improved methods to monitor and predict movement of the surface facility relative to the wells will improve the operation.

Even with all of the imagination that I can muster, I feel this brute force approach will not allow development of reserves under the polar ice cap. Thus, this system cannot be universally applicable.

Subsea development of the ''subsea city'' approach, involving both seafloor and off-seafloor habitats, has been proposed. The problems for both concepts are enormous. Consider transporting and maintaining 20 to 50 people at a remote subsea location with need to supply daily their food and life support, as well as several hundred tons of drilling supplies, and perhaps 20,000 horsepower of energy to a subsea facility. Beyond this, consider the technical problems of passing a rotting drillpipe or even a flexible pipe to operate a downhole drill through the wall of a habitat hull. I lack the imagination to consider even seriously this as a viable alternative.

Even if the equipment and logistics problems can be solved, improved well control methods are required. Contingency plans for drilling a relief well in case of a blowout would have to be incorporated into the system design.

Production systems installed below the seabed are feasible, assuming suitable rock formations for tunneling can be found between the location of the surface facility and the hydrocarbon reserves. The major problem is developing methods of exploration to determine in advance that the hydrocarbon reserves are large enough and prolific enough to justify the cost of the tunnel. Other very significant problems are route geology and contingency planning against loss of well control. Logistics problems are less with the sub-seabed system than with the surface or subsurface systems.

Improved methods of selecting a tunneling route are needed. New tunneling machines, using the latest rock mechanics and hardware reliability/maintainability methods, are required to allow effective tunneling for distances of 100 miles or more.

As with the subsea systems, well control methods must be made as certain as possible. Above water, an uncontrolled blowout can kill people and burn the rig or platform, very regrettable and serious events. Subsea or underground, an uncontrolled blowout could kill an entire staff of people, resulting in total loss of the production system and even possibly the reserves, not to mention the very serious environmental damage that could occur.

Summary—Needed Technology

- at least exploration drilling methods—at any cost—to define Arctic reserves before permanent facilities are installed
- improved, absolutely foolproof, well control/contingency planning
- methods to keep floating surface production facilities on location in severe ice
- methods to install subsea well equipment on the seafloor under ice
- methods to handle long production risers to free the surface vessel stationkeeping requirements to practical levels
- improved subsea tunneling methods.

Iceberg Resisting Offshore Structures For The 21st Century

L.C. Zaleski-Zamenhof
C. G. Doris

Arctic Oil and Gas Workshop (B)

The development of Arctic engineering for the 21st century should be consistent with a simultaneous development of the design philosophy and technology of materials, from which future structures have to be built.

DESIGN PHILOSOPHY

First, the structures have to be safe; they have to be safe in the 21st century, as safe as our structures designed and built in the 20th century are. However, the engineer's approach to the structural safety concept is still progressing from a deterministic one yesterday, through the semi-probablistic of today, to an actual probabilistic approach tomorrow, thus enabling it to fit the reality better and better.

Deterministic safety coefficients, when applied on the strength of material, result in admissible stresses. A respect of these stresses had to present an apparent panacea against any structural collapse. Nevertheless, structures still collapse from time to time. Thus we are aware nowadays that a structure presenting a probability of failure equal to zero would involve an expense equal to infinite (Figure 1). Subsequently, the Euro-International Concrete Committee[1] is agreeing in its present Recommendations on a probability of failure not exceeding 10 exp-5 per year for concrete structures in general.

Figure 1.

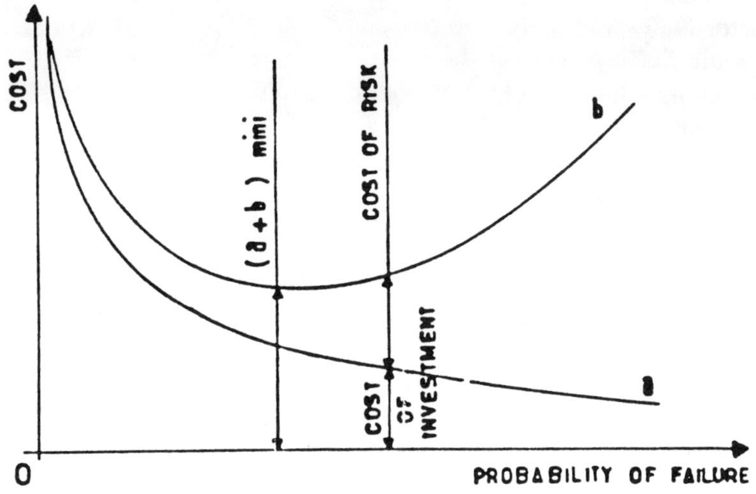

This means that among each hundred thousand structures of various kinds over the world, there would be one that could be damaged each year. However, the safety criteria become more severe for collapses endangering human lifes, resulting in heavy economic losses and or environmental disasters.

Therefore, instead of a unique safety coefficient, we should apply the so-called partial ones as the load and material factors; figures to be evaluated versus the reliability of assumed load events and of assumed material strength, respectively. This according to the presently recommended semi-probablistic analysis methods, which also govern the design of offshore structures.

Nevertheless, having tried to check the actual reliability[2] of several offshore structures as built and designed according to the partial safety factors required, I could find collapse probability figures varying from 10 exp-10 to 10 exp-8[3,4]; this means in the range from 1 to 100, but still lower than the expected ones. In practical terms, the checked structures have appeared overdimensioned, when for some other ones a too high probability of failure could eventually appear. Subsequently I had to conclude that in the 21st century (and certainly in several earlier cases), the actual probabilistic analysis, almost of level 2 (Figure 2), will have to be substituted to the present semi-probablistic level 1. This should occur at first with respect to structures designed to resist iceberg impacts and approach, as in offshore in general, and especially in the Arctic.

So many random parameters[5]:

- iceberg mass
- added mass coefficients } incoming kinetic energy
- initial berg velocity
- berg approach azimuth
- impact location
- berg shape
- geometry of impact area
- ice strength at given T and $\frac{d\epsilon}{dt}$
- ice temperature
- shape factor
- contact factor
- damping factors
- modal soil mass

as well independent and interdependent, actually govern iceberg impact events that an actual probabilistic approach, using a kind of Monte-Carlo simulation, appears to be the only one to be applied. Such an approach, once the appropriate reliability requirements are defined for economical, humanitary and ecological reasons, leads to a definition for a given accepted cost of an accepted probability of failure (as well as with respect to the global one as related to individual structural elements) and a probability of a progressive collapse (Figure 3). When comparing the iceberg impact events evaluation to the one applied with respect to earthquakes, the event corresponding to the first probability figure is then considered as the "design" event, while the second is assimilated to the exceptional or "more severe credible" event, for which some determined limited damages are considered.

Figure 2.

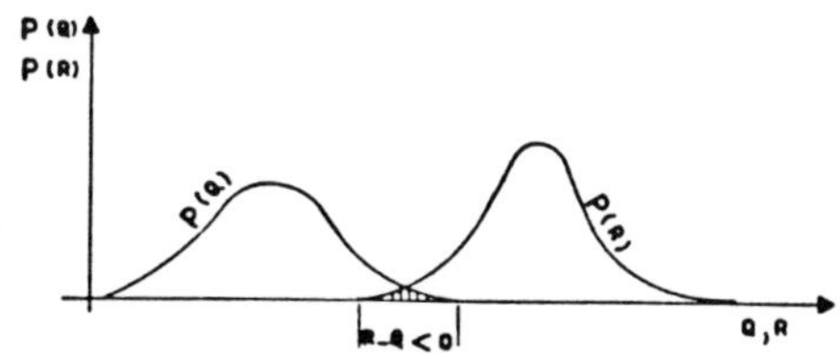

$\nu > 0$ when $R - Q > 0$
$\nu = 0$ when $R - Q = 0$
$\nu < 0$ when $R - Q < 0$

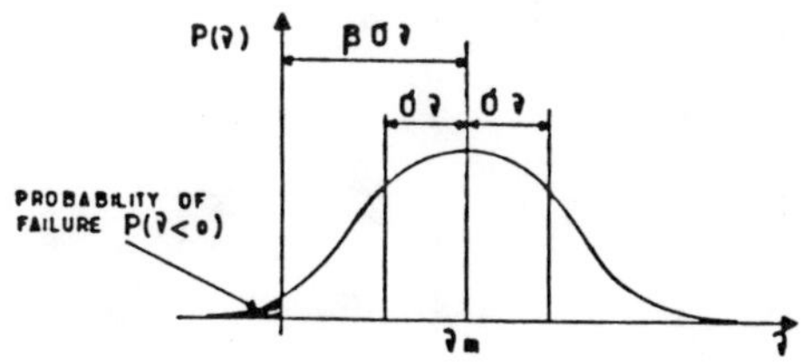

ex : $$\nu = \frac{1}{\alpha_Q V_Q + \alpha_R V_R} \text{Ln} \frac{R}{Q}$$

Where : $V_Q = \sigma_{Q/Q_m}$ $\quad V_R = \sigma_{R/R_m}$

Thus : $$P(\nu < 0) = \int_{-\infty}^{0} p(\nu)\, d\nu$$

$$\beta = \frac{\nu_m}{\sigma_\nu}$$

It appears important to a so-called ''balanced design'' of the structure not to over-dimension it. The impact forces actually are obtained from an energy absorption analysis; accounting step by step in the time domain for motions and deformations of the iceberg and of the structure, as well as for rigid body displacements as the elastoplasitc deformations of individual elements are concerned[6,7]:

Figure 3.

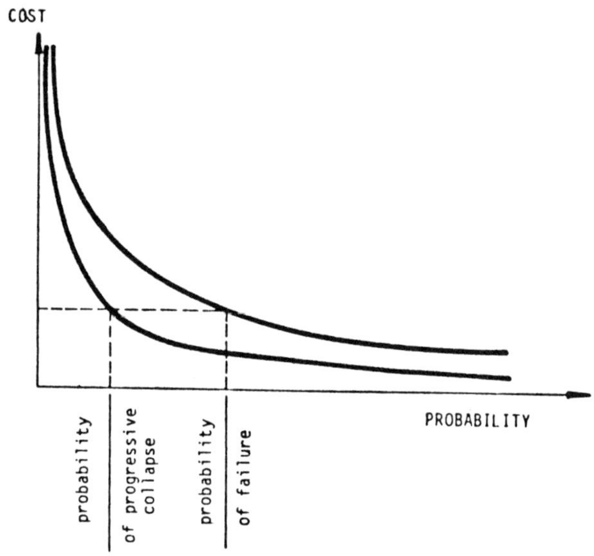

$$M_s \frac{d^2x_s}{dt^2} + A \frac{dx_s}{dt} + Kx_s = P\ (x_p)$$

$$M_b \frac{d^2x_i}{dt^2} + C \frac{dx_i}{dt} = -\ P(x_p)$$

$$x_p(t) = x_i - x_s$$

$$P(t) = I * C_m * C_k * f_c\ (T, \frac{d\epsilon}{dt}) * D * e$$

$$\int_{t_o}^{t_e} P(t)dx_i = M_B \frac{V_o^2}{2}$$

Subsequently, when accepting for the exceptional event a significant overall horizontal displacement of the structure (by sliding or by a deformation of a large mass of foundation soil jointly with elastoplastic deformations of the ''ice wall,'' i.e., the structural element hit directly by the iceberg), the resulting impact forces should not present much higher figures than the ones involved by the design event, thus not necessarily governing the design.

A question could then be raised: if, when hit by an iceberg, large displacements of the structure appear beneficial for impact energy absorption and result in decreasing the impact forces, why should we not substitute to the bottom-founded platforms any floating ones, with a conventional mooring system, able to absorb the impact energy through large deformations of mooring lines? It should be insisted that such a deformation, for a design impact, should be compatible with the one deemed acceptable, for example, for the risers. At the same time, a disconnection of risers could be foreseen for events ranging from the design to the exceptional one. It should be noted that for deeper waters, the same accepted tilt angle of a riser involves larger acceptable horizontal displacement of the floating structure. It could subsequently be concluded that:

- in ice-infested waters in general, and also as far structures designed to resist iceberg impacts are concerned, the choice between a floating and a bottom-founded platform may depend on the prevailing water depth;
- even in medium-depth range (about 100 meters), floating platforms could be used for exploration or for early production purposes. Such platforms actually present a possibility for reuse; furthermore, for an operational life on one site of no more than 1-2 years, the design event would involve a quite smaller iceberg impact as on a platform conceived to operate for some thirty years.

To complete a large range of Arctic exploration and production structures,[8] a concept of a fixed platform, offshore Newfoundland, designed to resist icebergs and to provide appropriate drilling, production and storage facilities, is presented in Figure 4.

Figure 4.

A concrete semi-submersible (Figure 5), derived from a Doris design,[9] could eventually serve to support drilling and/or early production facilities as well as in Arctic regions as in iceberg-infested waters.

Figure 5.

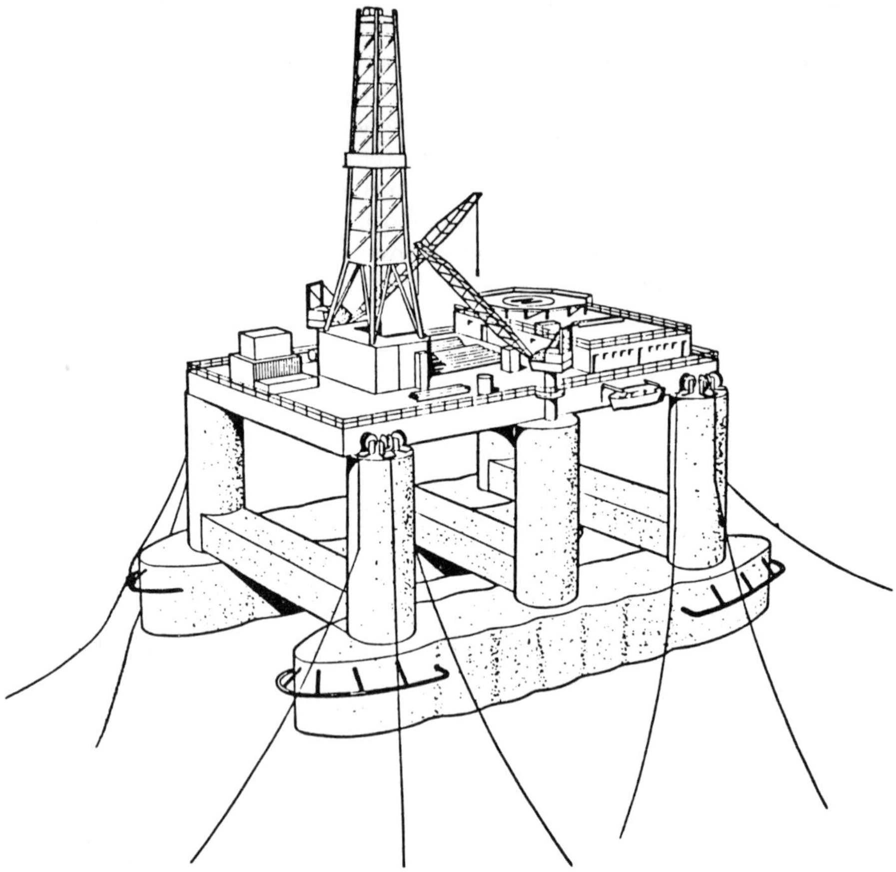

CONCRETE TECHNOLOGY DEVELOPMENT

As far as structural materials are concerned, this presentation has been limited to the concrete applications. Is this material, more than 100 years old, compatible with 21st century engineering?

I would say yes; this material is actually quite modern.[10] I still remember an epoch, when a compressive concrete strength of 15 to 20 MPa could be considered as correct. Today, a concrete strength 50 to 60 MPa is commonly required and obtained for offshore structures, and a figure of 100 to 120 MPa is considered, if such a one appears necessary. Furthermore, as concrete strength of 200 MPa and more is at present reached in some laboratories, we may assume such concrete could be available for building the platforms of the 21st century.

On the other hand, concrete ductility is still being improved, by appropriate reinforcement design, composite steel-concrete sandwich elements and "admixtures"—glass, carbon, aramide, steel or cast-iron fibers—to cast concrete material. Tests are showing that such material presents a resistance significantly improved against impacts as well as against abrasion—important points as far as Arctic structures are concerned. And, last but not least, such concrete should no longer be considered as brittle material since it reaches a tensile strength of 10 MPa!

Another old handicap of concrete is its low strength/density ratio but this too seems to disappear with the development of ultra high-strength concrete. Simultaneously, a lighweight structural concrete, presenting a compressive strength of 60-70 MPa for a density of 1.9 T/m3, also contributes to eliminating this handicap.

Therefore both the development of a design philosophy and the development of construction materials may be considered important factors promising an exciting future to Arctic engineering for the coming century.

REFERENCES

1. Comité Euro-International du Béton, "First Order Reliability Criteria (Level 2 Methods)," *Bulletin d'Information* N° 124/125-E, App. I, Paris, 1978.
2. Røren, E.M.Q. and O. Furnes. "Behaviour of Offshore Structures and Structural Design," BOSS '76, Trondheim, 1976.
3. Zaleski, L.C. "FIP Commission on Concrete Sea Structures—Risk Analysis," FIP Notes 95, Wexham Springs, 1981.
4. Le Tirant, P. and L.C. Zaleski-Zamenhof. "Evaluation de la Sécurité des Fondations Marines: Applications aux Ouvrages à Embase Poids"; 2nd International Symposium on Behaviour of Offshore Concrete Structures, ARBEM, Paris, 1982.
5. Deleuil, G. and L.C. Zaleski-Zamenhof. "A Methodology of Evaluation of Iceberg Loads on Fixed Offshore Structures," FIP/CPCI Symposium on Concrete Sea Structures in Arctic Regions, Calgary, 1984.
6. Croteau, P. "Dynamic Interactions Between Floating Ice and Offshore Structures," College of Engineering, University of California, Berkeley, 1983.
7. Sedillot, F. and L.C. Zaleski-Zamenhof. "Offshore Platforms in Arctic Area, Towards more Realistic Design Criteria," Third International Offshore Mechanics and Arctic Engineering Symposium, New Orleans, 1984.
8. Rojansky, M. "Arctic Exploration and Production Structures," Marine Technology Society *Journal*, Vol. 18 N°1, Washington, D.C. 1984.
9. Pham, G. and L.C. Zaleski-Zamenhof. "Des Semi-submersibles en Béton Précontraint—Une Nouvelle Génération des Plates-formes Flottantes," 2nd International Symposium on Behaviour of Offshore Concrete Structures, ARBEM, Paris, 1982.
10. ISSC 1985 Committee III.2., Non Ferrous and Composite Structures, Chapter 4: "Reinforced and Prestressed Concrete," 9th International Ship Structures Congress, Geneva, 1985 (proceedings to be published).

Petroleum Exploration on the National Petroleum Reserve in Alaska and Protection of the Environment

Max C. Brewer
U.S. Geological Survey

Environmental Considerations Workshop

The National Petroleum Reserve in Alaska (NPRA) comprises about 95,900 square kilometers of the tundra-covered area of the North Slope. Its northern boundary is the Arctic Ocean, including portions of the coastline of both the Chukchi and the Beaufort Seas. A little over one-half of its southern boundary follows the divide of the Brooks Mountain Range.

Except for a portion beneath Teshekpuk Lake, which is about 810 square kilometers in area, the entire Reserve is underlain by permafrost, ranging in depth (to the best of our present knowledge) from about 200 to 400 meters. The minimum temperature of the permafrost, below the depth of measurable annual change, is -10.7°C. The thickness of the active layer, depending on the material, topography, location within the Reserve, and the vegetative cover, ranges from 0.3 to 1.5 meters. The annual precipitation averages about 0.3 meters; almost 80 percent of that occurs in the form of snow. Thus, the area should be considered a polar semi-desert, in spite of the myriad of generally shallow lakes (1 to 3 meters) that dot the Coastal Plain area. Most of the Reserve, except that portion near the mountain front, is almost devoid of gravel, essential for construction purposes. However, approximately 13,000 square kilometers of the area is covered by mostly-stabilized sand dunes.

Exploration of the Petroleum Reserve for oil and gas potential generally has been concentrated into three time frames. The first was a geological reconnaissance, during the period 1923–36, by personnel from the U.S. Geological Survey. This was followed by the Navy's exploration program, involving geologic mapping, geophysical prospecting and the exploratory drilling of 36 generally shallow test wells and 44 core test wells during the period 1944–53. The most recent program occurred during the period 1974–81, when first the Navy and then the U.S. Geological Survey accomplished 21,800 kilometers of reflection seismic survey and drilled 28 exploratory and 10 development wells—the latter in the small, shallow natural gas fields near Barrow. Two of the exploratory wells were the deepest ever drilled in Alaska, approximating 6200 meters in depth.

The Navy's 1944–53 exploration program began under wartime conditions and even used some military equipment such as tracked landing vehicles for transport of men and equipment. It was a pioneering effort in which one of the goals was to see what equipment would work in the rigorous Arctic environment and what modifications might be required in exploratory techniques. The most important of the latter was learning [the knowledge] that successful overland travel must be a wintertime effort. The environmental protection ethic was not being emphasized at the time, nor was research advanced far enough concerning the impact of this type of operation on the Arctic environment. In retrospect two serious environmental impacts resulted. The first involved overland transportation, especially during the summer periods when, in order to obtain traction and to keep from high-centering vehicles, the "active layer" was bladed down to permafrost. In some instances, this resulted in erosion; in all cases it left ditches across the tundra that are expected to be visible longer than the old Roman roads in Britain have been. The second big environmental impact was that debris, especially empty fuel drums, was scattered across the tundra. The latter problem has been largely rectified, the debris gathered up and burned or buried, but at a very large financial cost.

One very positive result of that earlier exploration was that in 1947, the Navy introduced a program of research in Arctic environmental matters, including operating techniques. The research program continued on, at the Naval Arctic Research Laboratory at Barrow and, thus, by the time that oil was discovered at Prudhoe Bay, an impressive amount of knowledge had been accumulated. This included, information regarding casing collapse in wells; the thicknesses of gravel required along the Arctic Coast to bring the permafrost surface [frost line] up into the gravel fill and in turn, to prevent the deterioration of the underlying permafrost in all-season roads and airstrips; building research; fresh ice and sea ice research; and, a far greater appreciation of the physical and biological parameters to be found in the Arctic.

Oil development at Prudhoe Bay has generally been cited as being a model of task-accomplishing, where a decision has been made, in an environmentally acceptable manner. As a result, during the past few years, whenever a program is projected for the Arctic area, there has been a tendency for both engineers and many environmentalists to want to use methods and parameters shown to work at Prudhoe and, at the same time, to leave a neat and tidy appearance. This has some environmental advantages, but also has many environmental disadvantages for an oil and gas exploration program.

Prudhoe Bay is a development complex with a transportation system. The roads, pads, airstrips, and other facilities are permanent, at least for 40 or 50 years. In the Arctic some of them will be visible for hundreds of years. However, in an oil and gas exploration program, one does not want permanence, at least not until after a discovery is made. Impermanence not only provides greater environmental protection but also is economically less costly in resources. As a result in an exploration program, one wants to "tiptoe" into an area, run the seismic surveys, drill any exploratory wells deemed desirable, and then to carefully move out leaving no tracks or other evidence of activity.

The Navy/U.S. Geological Survey exploration program of 1974–1981 tried to follow the approach of leaving as few tracks of their exploration as possible. All seismic surveys were run during the winter-spring period when the snow cover was greatest, when overland transportation would be most efficient and when the least numbers of wildlife were present. Exploratory drilling was confined to the winter season, except where deep wells in excess of about 3500 meters were involved. Beyond that depth, the operator would be caught with the trade-offs between constructing all-season road and airstrips and the environmental impacts and costs of an operation extending over several winters. In addition, the operator would be faced with a lack of surety in completing a well, without losing it, where drilling would have been suspended several times during the summer seasons.

Lake ice was used for aircraft landing surfaces wherever possible. In two instances, airstrips to accommodate Hercules C-130 aircraft were constructed directly over a flat tundra expanse. In one instance an ice airstrip was constructed in the same place the following year. After the second season, it was extremely difficult to find any evidence of the 30-centimeter thick airstrips having ever been constructed. In the case of three deep wells, high-strength insulation was used in airstrips in order to reduce the thickness of gravel required to bring the permafrost surface [frost line] into the base of the fill.

Ice roads were constructed wherever heavy traffic, such as 20-cubic meter dump trucks, might be making repeated trips. These were built to a minimum thickness of 15 centimeters before heavy traffic was allowed. The longest ice road constructed was 60 kilometers. Water was periodically applied to keep the road in good repair and to maintain a smooth running surface. By the time 85,000 cubic meters of gravel had been hauled over this road, at speeds of up to 80 kilometers per hour, the built-up ice road had an average thickness of about 45 centimeters. Almost no damage, not even a ''green'' trail (resulting from a compressed tundra mat which affords plants better access to moisture) or a trail discernible from a low-flying helicopter, was evident the following summer. Snow roads, per se, were not constructed because of the generally insufficient amount of snow available and because, whenever snow is used it must be scraped up. This requires blading with tractor-type equipment and, no matter how careful the operator is, considerable tundra damage usually ensues.

While 1.5- to 1.7-meter thick gravel drilling pads bring the frost line into the base of the pads in northern Alaska, a pad of such thickness will be quite visible in the long term, will be well-drained and difficult to revegetate because of a lack of sustained moisture at root level. A thick pad, however, is not necessary to support a winter-only drilling operation. A thin pad (0.5 meter thick) frozen during the winter months, satisfactorily supports a winter-only drilling operation; requires no added material to be hauled to the site since sufficient material is available after digging a reserve (mud) pit; and retains sufficient moisture to enable grasses to become established. It is, therefore, easier to revegetate and can rapidly recede into the surrounding landscape. These thin pads were used on the National Petroleum Reserve in Alaska (NPRA) during the latter portion of the exploration program and proved to be more satisfactory than the thick pads from both the engineering and environmental standpoints. They also eliminated the need to open material sites, and thus the need for hauling heavy loads of gravel. In addition, they were more economical to prepare.

The recent exploration program on the NPRA received few complaints from environmental groups and since then, its completion has occasionally been cited by such groups as the model way to accomplish an exploratory program in the Arctic. The program also was completed approximately 12 percent under its budgeted funding.

Future Developments In The Soviet Arctic Marine Transportation System

Lt. Cdr. Lawson W. Brigham
United States Coast Guard

Transportation Workshop

Any discussion of the Soviet Arctic marine transportation system must pause to reflect on the extraordinary achievements the Soviet Union has made during the past thirty years. Table 1 provides a glimpse of several accomplishments which hold great significance for the marine world with regard to transportation in the Arctic Ocean. Of interest is the fact that the administration of the Northern Sea Route has been particularly bold and innovative in applying nuclear power to icebreakers, has

Table 1. Soviet Arctic Marine Transportation System: Accomplishments.

- World's First Experimental Ice Tank For Ship Research (1955)
- First Nuclear Powered Icebreaker (1959)
- Development of the World's Most Powerful Icebreakers
- First Transit to the North Pole by a Surface Vessel (1977)
- First Icebreaker Escort of a Freighter Across the Arctic Basin (1978)
- Routine Operations of Icebreaking Cargo Vessels (SA-15 Class)
- Employment of Specialized Icebreakers (Shallow-draft Polar & River Icebreakers)
- Substantial Ice Piloting Expertise Due to Extensive, Long-term Efforts
- Achievement of a Nearly Year-round Navigation Season on the Northern Sea Route
- Successful Application of New Technology (spoon-shaped bows, ice breaking-ice removing devices, cp propellers, bubblers, quadruple-screw propulsion systems)

conducted highly successful Arctic Ocean transits and has utilized technically advanced equipment. The full measure of the Northern Sea Route operation is often difficult to comprehend in North America.

In keeping with the spirit of the Symposium, this paper will provide a cursory look at the future of the Northern Sea Route system. Its primary focus will be on the year 2000 and beyond. Although the subject is highly speculative, the paper will attempt to review the planning that must take place for Soviet polar icebreaker replacement, review the various transport modes that are feasible in the Soviet Arctic and comment on future nuclear power developments.

USSR POLAR ICEBREAKER FLEET BY 2000

One way of defining the present composition of the world fleet of polar icebreakers is to include only those ships strictly designed for heavy icebreaking that are capable of independent Arctic operations in multi-year ice. This definition limits the fleet total to 36 vessels. Nearterm additions are believed to be *Rossiya,* a fourth *Arktika* class ship, a fourth ship of the Canadian *Radisson* class and perhaps two Soviet, shallow-draft polar icebreakers to be built by Wartsila of Finland. Thus, by the late 1980s the USSR will be operating 19 heavy icebreakers—approximately fifty percent of the world polar icebreaker fleet. Notable, minimum characteristics of the Soviet vessels will be a 120-meter length, 15,000-ton displacement and 22,000-shaft horsepower.

Table 2 indicates a future age profile of the Soviet fleet. The two new nuclear vessels listed are of the *Arktika* class (*Rossiya* and a fourth ship of the class). As of the date of the Conference, negotiations continue between the USSR and Wartsila for the delivery of shallow-draft polar icebreakers of the *Taymyr* class. Although it is unclear

Table 2. Average Ages of the Soviet Polar Icebreaker Classes by the Years 2000–2010.

Class	Number Vessels In Class	Average Age 2000	Average Age 2010
BREZHNEV/SIBIR	2	24	34
NEW NUCLEARS	2	14	24
LENIN	1	41	51
YERMAK	3	25	35
MOSKVA	5	31[1]	41
KAPITAN DRANITSYN	2	20	30
KAPITAN SOROKIN	2	23	33

Note: 1 Youngest Vessel in the Class

when these ships will be completed, they would be relatively young by 2000. Significantly, Table 2 shows that 15 heavy icebreakers will be 20 years or older by 2000. The *Lenin* and five ships of the *Moskva* class will be beyond or at retirement age. Plans to replace the workhorse, diesel-electric powered *Yermak* class (36,000 shp and 128m in length) will no doubt be well underway. Preliminary plans should have been undertaken for new ships to replace *Brezhnev, Sibir* and both *Kapitan Sorokin* and *Kapitan Dranitsyn* classes.

The capital costs of such a ship replacement program will be extraordinary considering the scope of the fleet's block obsolescence. The Ministry of the Merchant Fleet is apparently convinced that larger, more powerful icebreakers are necessary for the majority of ice escort duties. This trend of increasing size will be reflected in rapidly increasing costs. It is anticipated that the highly successful SA-15 icebreaking cargo vessel will have an impact on the size, power and type of future polar icebreakers. Finally, a formidable challenge to be met will be the appropriate timing for replacement of the fleet. Considering the lengthy acquisition process and the likely desire to spread out the age of the fleet, a carefully orchestrated replacement program will be mandatory.

MODES OF ARCTIC MARINE TRANSPORT

Three general modes of marine transport can be identified for ice-covered waters—submarine, icebreaker escort of ice-strengthened cargo ships, and independent operation of icebreaking cargo ships. It is important to analyze which of the three are potentially attractive, technologically feasible and operationally sound for use in the Soviet Arctic. Because considerable attention has been devoted to potential cargo submarine operations in the North American Arctic, two questions are raised with regard to their use in the Soviet Arctic. How feasible is such a system along the entire length of the Northern Sea Route? And, is it likely that the USSR would employ its own or foreign cargo submarines by the 21st century? Figure 1 hints at a recent Soviet development of a semi-submersible icebreaker.[1] However, the technological difficulties associated with the icebreaking process (around the protruding pylon) and maintenance of a semi-submerged state in the ice (with neutral buoyancy) would appear unresolved.

Although large submarine operations in the Barents Sea are highly feasible due to sufficient water depths, a significant drawback would be the short transits to the Soviet mainland. If oil and gas could be transported to European or North American markets, cargo submarines might be utilized. In the Kara Sea the extended continental shelf would require offshore terminals and underwater pipelines. Development of a submarine system for the Laptev and East Siberian Seas is considered highly unlikely due to their shallow, broad shelves and the anticipated lack of development by the year 2000. An overriding consideration might be that the Soviet Union has already invested heavily in a surface fleet for the Northern Sea Route. The technological risks, safety concerns and high cost of a subsurface system would appear to be insurmountable obstacles. Further, it would not appear possible that foreign commercial submarines would ever be allowed to operate so near to the Soviet mainland.

Table 3 is a summary matrix that compares the three modes of transport to the various seas that lie along the Northern Sea Route. Recent operations have proven

Figure 1. Revolutionary Soviet Design of a Semi-Submersible Icebreaker.

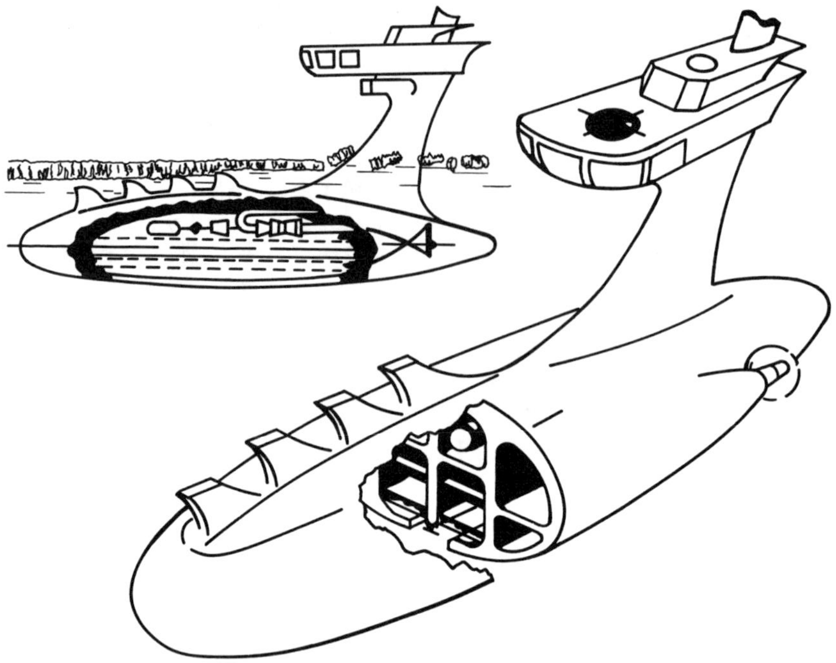

that icebreaker escort of cargo vessels is technologically and operationally feasible across the entire breadth of the Route. However, severe conditions in the east can appear unexpectedly and make convoy operations difficult if not hazardous. Such was the case in October, 1983, when a fleet of fifty Soviet resupply ships and several polar icebreakers became locked in severe ice conditions in the East Siberian and Chukchi Seas. Northerly winds and subfreezing temperatures made conditions extremely difficult. After three weeks of effort, 13 polar icebreakers were able to free the convoy, but not without the loss of one ressupply vessel and extensive damage to over half the ships involved.

In the Barents Sea, icebreaker escort most probably will be considered an excessive and unnecessary capability since, even today, the SA-15 icebreaking cargo ships can negotiate independently that region's ice fields with little problem. Future nuclear, lighter-aboard-ship (LASH) vessels should also possess a similar capability. By the year 2000 economics and efficiency of operations will point to such a transport mode throughout the Barents Sea. It is also probable that by 2000 the Administration of the Northern Sea Route may not have to depend on continuous and extensive icebreaker escort in the Kara Sea. Polar icebreakers have recently kept the southwest Kara Sea near the Yenisey River open for ten months of the year.[2] Coupled with proper timing and enhanced ice information, the icebreaking capability and large cargo capacity of the new ships make it highly feasible that Kara Sea operations can be conducted without icebreaker escort in the 21st century. As to the severity of ice conditions in the Laptev and East Siberian Seas, independent icebreaking cargo ship operations are not envisioned by the year 2000.

Table 3. Feasibility Matrix for the Soviet Arctic by the Year 2000.

Mode of Marine Transport

Seas	Submarine	Icebreaker Escort	Independent Cargo Ship Ops
Barents	Yes	Yes[1]	Yes
Kara	No[2]	Yes	Probable
Laptev/ East Siberian	No[3]	Yes	No[4]

Notes: 1 Not as economical as independent cargo ship ops.
2 Shallow waters and the requirement for offshore terminals.
3 Shallow waters and undeveloped by 2000.
4 Ice conditions difficult and undeveloped by 2000.

NUCLEAR POWER DEVELOPMENTS

The key to any future activities, related to extending the navigation season along the Northern Sea Route, will be the continued application of nuclear power to various types of icebreaking vessels. The fact that the *Arktika* class has been highly successful no doubt contributed to the decision to build *Rossiya* and a fourth, as yet unnamed ship of the class. However, these vessels are deep draft (10.6 meters), and experience with the *Kapitan Sorokin* and *Kapitan Dranitsyn* classes has shown that an 8-meter draft is much preferred for working the mouth of the Siberian rivers. The Soviet requirements for greater power, shallow-draft design and enhanced endurance gave rise to preliminary plans for nuclear powered, shallow-draft polar icebreakers. It is believed that Wartsila of Finland will build the ships and deliver them to the Soviet Union for installation of nuclear reactors. The technical difficulties of such a joint arrangement appear surmountable.

In all probability, a fleet of icebreaking LASH transport vessels will be sailing the length of the Northern Sea Route by the year 2000. Figure 2 depicts some of the significant features of such a ship that will be characterized by the unlimited endurance of nuclear power as it plows along the Siberian rivers.[3] These ships will be totally independent of port facilities—a necessary feature along most of the undeveloped Siberian

Figure 2. Soviet Nuclear Icebreaking Lighter Carrier.

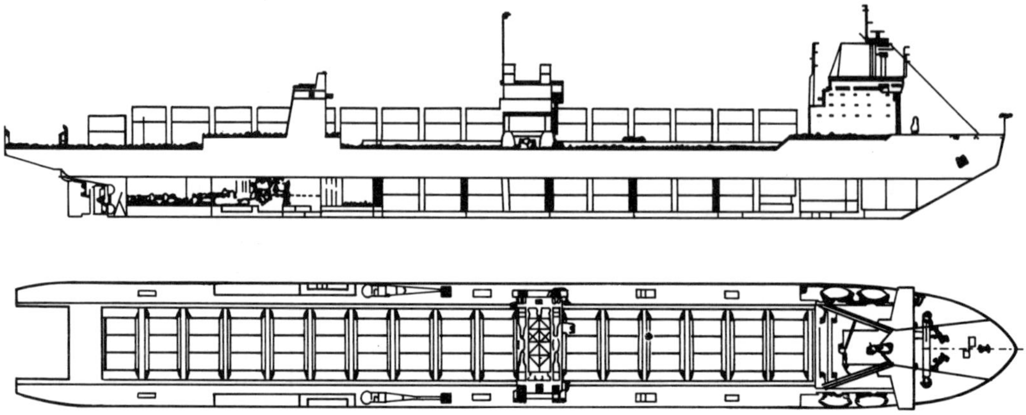

(Length Overall: 260m: Beam: 32m; Draft: 10.6 to 11.8 m; Displacement: 61,000 Tons; Speed: 20 Kts)

coast. Speculation can be advanced that coupling these vessels in convoy with the *Arktika* class or new shallow-draft vessels will produce a virtually unstoppable team. In the 1970s, the average speed of a convoy along the Route was 3 to 4 knots.[4] The nuclear combination would be capable of attaining almost three times that speed, under normal ice conditions.

FORECAST FOR 2000 AND CONCLUSIONS

While it is difficult to predict the Soviet Union's activities along the Northern Sea Route, several trends make reasonable speculation possible. Normally, the decision of another nation, with regard to new icebreaker construction, would have no bearing on the future of Soviet development and presence in the Arctic. However, if Canada finally decides to build the *Polar 8* ship with 100,000-shaft horsepower, the Soviet Union would be encouraged to forge ahead with an icebreaker of gigantic proportions, perhaps 150,000-shaft horsepower. Block obsolescence of several large polar icebreakers in the early 21st century would also influence a Soviet decision to go ahead with a ship capable of extended operations anywhere and anytime in the Arctic basin.

The Northern Sea Route will be sailed by fewer, but larger and more ice-capable tanker and cargo vessels. Nuclear power and LASH concepts will continue to be coupled in the development of modern transport ships. There will also be a routine application of nuclear power to shallow-draft polar icebreakers by the year 2000. In this regard the Finnish shipbuilder Wartsila will in all probability have completed the hull and propulsion for a nuclear powered polar icebreaker for the Soviet fleet. There is little doubt, though, that the Soviet Union will await Western development of tanker-cargo submarines for Arctic operations. Geography and an already heavy investment in a surface fleet preclude such a drastic change in direction.

Improvements along the Northern Sea Route will be made in the operations infrastructure, particularly in the area of all-weather ice information. There will be expanded use of vertical lift aircraft aboard both polar icebreakers and icebreaking cargo vessels. For shallow river operations, new channel clearing devices will be developed and improved and integrated tug-barge combinations may be employed.

In conclusion the Soviet Arctic marine transportation system developed during the past fifty years is an unprecedented polar achievement. The USSR's continued investment in a broad range of vessels and support equipment can only indicate a linkage to its future domestic economy. Increased resupply efforts, the movement of Siberian raw materials out of the northern regions, and the development of oil and gas will all require an enhanced capability. It is clear that the Soviet Union is embarked on an improvement program for the Northern Sea Route operation in order to support development of the entire Siberian industrial region.

REFERENCES

1. Prescott, J. USSR Designs Revolutionary Semi-Submersible Icebreaker, *Lloyds List*, London, February 2, 1984.
2. Armstrong, T. The Northern Sea Route Today. *Cold Regions Science and Technology*, 7: 251-257, 1983.
3. Watson, G.G. The Soviet Nuclear Icebreaking Lighter Carrier. *Marine Engineers Review*, 18-20, January, 1984.
4. Maksutov. D.D. (translation by G.G. Watson). Soviet Icebreaker Construction: Its Present Position and Future Prospects. Transactions of the Arctic and Antarctic Scientific Research Institute, Leningrad, 376: 14-21, 1981.

SUBMARINES IN PERSPECTIVE

Larry Jacobsen
General Dynamics

Transportation Workshop

Since there are no submarine tankers today, I don't think I should be challenged on where I think they will be in the future. And I'm not inhibited by what can't be done now.

Seventy years have elapsed since two German cargo submarines transported cargo from Baltimore, Newport, and New London to Bremerhaven, albeit, a covert transit. Today over 800 submarines are operational, of these over 350 are nuclear propelled. These submarines (mobile submerged platforms for weapons systems) are operating under the flags of some 37 nations of the world. Historically, submarines have been employed in covert transport of people, weapons, fuel, and precious and strategic materials.

Of the four nations possessing nuclear powered submarines, three have carried out "scientific" operations in the Arctic, transiting under ice at will and routinely surfacing in open and ice covered polyneas. Even at the geographical north pole as recently as the winter of 1982–83, two U.S. Navy submarines, SSN 639 and SSN 648, surfaced. The 648 surfaced after 43 days submerged. The last reported time when two surfaced at the pole was in 1962.

Technical credibility of submarine operations in the Arctic is not an issue now or in the future *and* has the potential to become the least disruptive to the Arctic environment of any transportation system.

The quest for modern cargo submarines dates from 1954. These concepts were given impetus by the development of marine nuclear power plants with full recognition of the potential on specific routes, in particular, under ice. The efforts have ranged from concepts to fully developed designs. They exposed submarines sized from 20,000 DWT oil carriers to 125,000 cu. m. LNG carriers. Cargo submarine concepts have originated in the United States, Japan, United Kingdom, and Canada. The common thread is the projected hydrocarbon yield from Arctic regions and the submarine's potential for transport under ice. Perceiving that the size of oil and gas finds, predict demand, and Arctic costs will not deter development in the North, a role for submarine tankers is foreseen.

If one accepts the scenarios of others at a high supply peak of 1,000,000 BPD and a 200,000 BPD as a conservative supply from either the Beaufort Sea Region or the Canadian Island Regions, a marine transport system will emerge with a role for the submarine. At this juncture, I do not foresee a future role for the submarine in the Beaufort or Chukchi Seas. There is a basis, however, to predict that submarine transport may have a prime role as a link in the inter-island transfer of oil and gas among the Canadian Queen Elizabeth Island Group and the mainland by the year 2010.

Following are four route scenarios which might well employ submarines in shuttle tanker service over relatively short routes:

1. From production in the Queen Elizabeth Island Group to Herchel Island; pipeline to Prudhoe Bay
2. From the Queen Elizabeth Islands via Peel Sound to terminal on Boothia Peninsula; pipeline to Port Arthur via Churchill
3. Inter-island transport (shuttle) to cross island pipeline to a most favorable surface tanker terminal (probably east of the longitude at Resoute)
4. Inter-island transport to a combined submarine/surface tanker floating transfer terminal at Radstock Bay or other rendezvous.

Indeed, these suggest that intermodal transport will occur in the Arctic. The submarine, then, is not projected to displace pipelines, surface ships, aircraft or other vehicles in the future—only to be employed in concert with these transport methods. Its role will be as a transport link for ice covered route segments unsuited to other transport modes.

To foster a submarine transport system, a favorable socio-economic climate must exist. I defer prognosis on investment, rate of return, or required freight rate in the year 2010. But I contend that maintaining a secure source of hydrocarbons in a world of volatile politics and economics can become an overriding driver and not an economic liability or lack of it. Submarine transport as a subsystem will be integrated with other transport subsystems to complete a total delivery system. As such, the submarine will be complementing, not competing.

Submarine tankers in the Arctic Shuttle Service are projected to be sized on the order of 40,000 cargo DWT tons or less. Some configuration evolutions will occur in the ensuing 20 to 30 years. But submarine transports will not vary significantly from those concepts which have been proposed.

Developmental areas which must be encompassed by Arctic engineering technology abound. For an Arctic transportation system to be put in place, a major terminal engineering feat must be accomplished, particularly for the submarine. It is not too early to start development of concepts. Two such terminal concepts for the submarine are:

1. open water harbors
2. multi-function floating structures (floating hangars) within which a submarine could surface.

Submarine propulsion engineering efforts can be expected to be undertaken to advance the art:

- pump jet propulsion
- fuel cell power
- gas cooled nuclear reactor power
- ducted propellers and reduced control surface.

Further, engineering efforts need to address concepts which can add flexibility of purpose to the submarine in the Arctic, such as:

- a long, topside, ice-strengthened fairwater, housing a command and control module which can be released as an escape submersible
- a tri-pressure hull ship section with the center hull pivoted at one end to permit tip up to the surface for logistic and cargo transfer.

I will admit that my crystal ball is now lined and wrinkled from overuse. However, in the event that 200,000 BPD or less is the high peak supply, permit me one venture—

that before the ice crystals clear, a hard look at containerized liquid cargo concepts will be undertaken.

Unit cargo transport is compatible with the submarine, surface ship, aircraft, train, and truck modes of transportation. A combination of these modes, transporting containerized liquid cargos, may well be found to offer acceptable solutions to many environmental issues of today. Twenty years hence, the environmentalist movement of the 70s and 80s can be predicted to have gained significant strength.

A challenge to the Arctic subsea transport engineers is to bring about solutions to the technological risks to the submarine transport system. There are risks until concepts are reduced to engineered solutions.

- crew escape and rescue
- management of operations (particularly nuclear)
- submerged cargo transfer
- docking interfaces
- floating multi-purpose terminals with submarine ''moon pools.''

I conclude that there is a role for the commercial cargo submarine in the Arctic. I have not addressed work submarines which are needed in the near future for research and will be usefully employed in construction support and maintenence of subsea structures and equipments.

THE TREND OF DEVELOPMENT OF STEELS USED IN THE ARCTIC OCEAN FIELD

Kametaro Itoh
Nippon Steel Corporation

Arctic Materials Workshop

Since ancient times iron has contributed to the development of the civilization of mankind. It may be said that the techniques for producing large amounts of molten steel by the blast furnace, LD converter and open hearth processes in the 18th and 19th centuries, as well as those for producing large amounts of light-gage sheets in the 1920s to 1930s and those for manufacturing large-scale welded structures in the 1940s and beyond, have been historic technological innovations worthy of special mention. Incidentally, these technologies are closely related to the use of energy. Today, in the Arctic Circle—the last exploitation area of natural energy—steel is expected to contribute to the energy exploitation of the 21st century to a greater extent than in past centuries. As shown in Table 1, steel structures have features superior to those of structures made of other materials that compete with steel. These features meet all the requirements that structures must meet in the severe environment of the Arctic Circle—safety, short construction term and economy. This report is made based on these features of steel structures.

Table 1. Merits for Steel and Steel Structures.

1. Voluntary control and combination of strength and toughness
2. A lot of properties data for ultimate design
3. Reliable quality control and quality assurance
4. Short manufacturing time (Suppliable of mass material, Fabrication system of large welding structures)
5. Reuse of steel structures, Recycling of steel material
6. Economy (Specific strength, Specific price)

First, this report describes trends in drilling production and transportation systems used for the energy exploitation in the Arctic that we can foresee. Secondly, it summarizes needs produced by the trends for properties which steels are required to provide.

Thirdly, it shows trends in the development of steels conducted to meet the needs by referring to the present situation. Future trends in production techniques of steels also are shown at the same time. Finally, it refers to welding, fracture mechanics and corrosion protection as software necessary for converting steel products into merchandise.

DEFINITION AND FEATURES OF THE ARCTIC CIRCLE

To foresee future subjects of steels and steel structures, it is necessary to know in what environment they are used. Geographically, the Arctic Circle is the area of the globe at lat. 66°30′ N. and northward. This definition, however, cannot unequivocally show the environment in which structures are used. The classification by E. Bergsager, which was shown in relation to energy exploitation districts, seems to be appropriate. This classification is as follows:

- areas north of Norway (at lat. 70° N. and Northward)
- areas along the USSR (Eastern Barents Sea, Kara Sea, and Eastern Siberian Sea)
- Alaska (Bering Sea and Beaufort Sea)
- Canadian Arctic (Canadian Arctic Islands, Beaufort Sea and Baffin Bay)
- areas along Greenland.

As is well known, these areas have the following features:

- distant from places of consumption
- cold climate (minimum service temperature: -50 to -60°C)
- strong winds and snow
- fast ocean currents and high waves
- existence of ice in various conditions (refer to Figures 1, 2 and 3.)
- existence of permafrost.

Figure 1. Mid-Winter Ice Conditions.

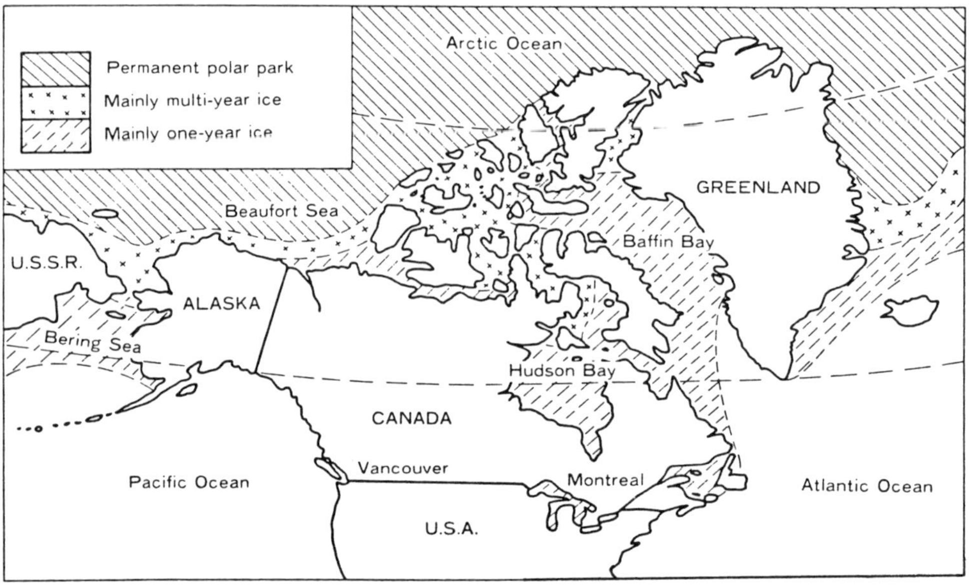

Source: A Review of Ship Research in Ice, RINA, 1978.

Figure 2. Mid-Summer Ice Conditions.

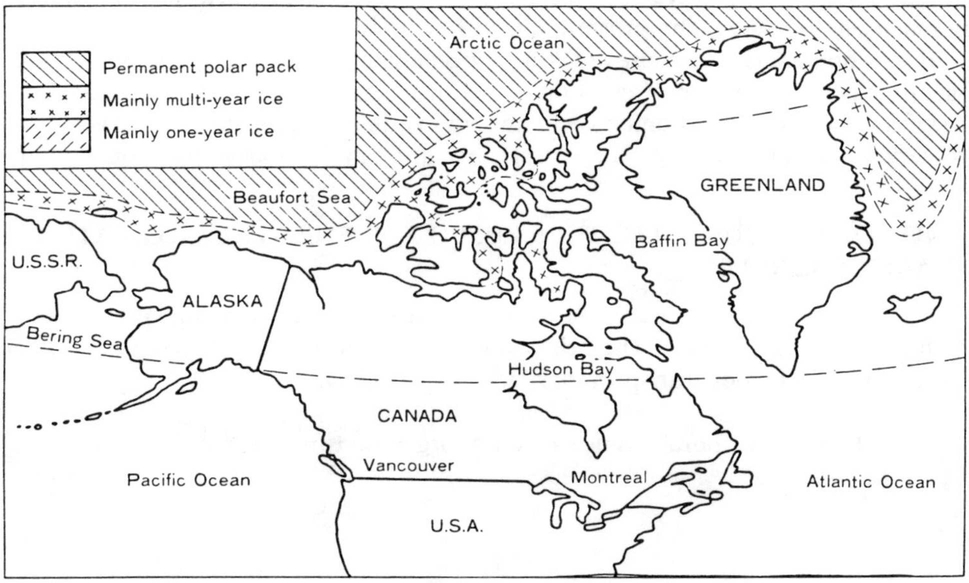

Source: A Review of Ship Research in Ice, RINA, 1978.

Figure 3. Spring Ice Conditions of the Canadian Beaufort Sea.

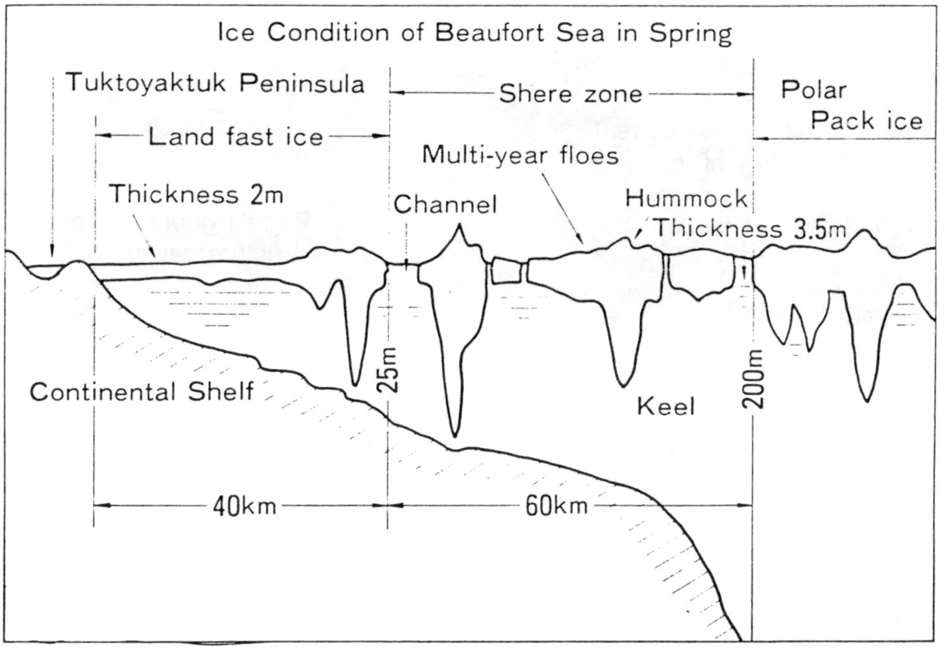

Source: '' **Source:** A Review of a Document from Dome Petroleum Co.

Water depths in the energy exploitation areas along the coast of the Arctic Ocean are relatively small. However, in other areas, such as those north of Norway, they may reach 300 to 500 m. The present author has no information that well depths and oil and gas compositions in these areas are especially different from those in other areas. From the basic mechanism of creation of oil and gas, however, they may be judged to be within the severest conditions ever known in other areas. In addition, it is naturally necessary to take the existence of the vast permafrost in these areas into consideration.

TRENDS IN VARIOUS STRUCTURES USED FOR ENERGY EXPLOITATION IN THE ARCTIC CIRCLE

This section describes future trends in structures used for drilling and production and ships for transportation. The brief outline is shown in Table 2. Trends in oil and gas wells and means of transportation are also dealt with.

Table 2. General Trends of Offshore Structures in the Arctic.

	→	
1. Economic competitiveness (Far from industrial areas)		1. Huge structure
2. Severe natural environment – low temperature – strong wind & snow – high wave, fast ocean current – variety of ice		2. Long life for operation
3. Pollution protection		3. Perfect guarantee of structure safety

Tables 3a and 3b show types of drilling and production structures, respectively, presently in use or under development. From the above-mentioned features of the oil and gas exploitation areas in the Arctic Circle, it is predicted that the following types will be in the mainstream of drilling and production structures used in the Arctic Circle.

Drilling rigs

The artificial-island type, which is very resistant to ice, will mainly be used in the future in sea zones of relatively shallow water along the coast of the Arctic Ocean. In this case, the movable caisson type will progress more than the gravel type. Steel structures, hybrid structures of steel and concrete and concrete structures are competing with each other for the caisson type. However, the more intensive development

Table 3a. Evaluation and Future in Each Type of Drilling.

Type	Operating Depth	Evaluation			Future
		for Ice	for Wave	all seasons operation	
1. Artificial island Gravel	Shallow Sea	◎	○	○	—
Caisson	Shallow Sea	◎	○	○	Main type for shallow arctic sea
2. Landing Gravity	Shallow Sea	△	△	△	—
Jack up	Medium Depth Sea	△	◎	△	—
3. Floating Ship or barge	Deep Sea	○	◎	△	—
Semi-sub	Deep Sea	○	◎	△	Main type for deep sea

◎ Very strong ○ Strong △ Weak

Table 3b. Evaluation and Future in Each Type of Production Platform.

Type	Operating Depth	Evaluation				Future
		for Ice	for Wave	For soft foundation	Technical completeness	
1. Artificial island	Shallow Sea	◎	○	○	◎	To be used continuously around shallow, offshore in the Arctic Ocean
2. Jacket	Medium Depth	○	◎	○	◎	
3. Gravity	Medium Depth	○	○	△	◎	To be limited in the sea with solid foundation
4. Floating	(Temporary)	○	○	◎	◎	
5. High-Compliance Guide-Tower	Deep Sea	△	○	◎	○	} To be used practically in the other sea area
TLP		△	○	◎	○	
6. S.P.S.	Deep Sea	◎	◎	◎	△	Will develop in the deep sea

◎ Excellent ○ Good △ No

of the types mainly using steel will be expected from the above-mentioned advantages of steel structures.

In sea zones of larger depths, off the coasts of the northern part of Norway, the Bering Sea and the Arctic Ocean, it is expected that drilling rigs will be capable of operating during all seasons in order to lengthen the exploitation period. For this purpose, new ideas for eliminating the effect of ice are desired.

Structures for production

Those structures are expected to operate for 20 to 40 years. Therefore, production structures are required to provide complete safety under the service conditions of the Arctic Circle.

In the shallow sea zone of the Arctic Ocean, along the coasts of Alaska and Canada, the artificial-island type is mostly used at present; this tendency will continue in the future. Fixed jackets of steel or concrete construction are used mainly in sea zones of relatively large depth, such as in the Norwegian and Bering Seas, etc.

If subsea production systems (SPS's) can assure complete safety, they seem to be ideal systems capable of completely eliminating the effects of ice, waves, ocean currents and temperatures. This type of production system might be put into practical use in the Arctic Circle through improvement in the reliability of control systems, sensors and subsea structures after the service experience in zones. For it is there that the climate is milder and the surrounding conditions more favorable.

Trends in Oil and Gas Wells

As is well known, the following techniques are carried out in oil and gas wells of the world:

- development of deep wells (that is, exploitation of high-temperature gases and oils)
- exploitation of oils and gases with sour compositions
- exploitation of mixed gases
- directional drilling
- positive application of secondary and tertiary recovery (that is, extension of well life).

Table 4 lists these trends as well as related requirements. It might be thought that similar developments will also be carried out in the Arctic Circle. In view of the pursuit of economic efficiency, the last two techniques mentioned above will be more thoroughly developed and applied in the Arctic Ocean than in other areas.

Means of Transportation

The economic competitiveness of oils and gases produced in the Arctic Circle is strongly influenced by whether the most economic transportation systems to places of consumption can be obtained or not. Table 5 gives means of transportation presently proposed or in practical use.

Because energy produced in the Arctic Circle is of vast amounts, pipeline systems are considered to be economically most favorable. The adoption of huge tankers also must be advantageous because energy is produced in the ocean. However, these two have many problems to solve due to the existence of ice in the Arctic Circle. The

Table 4. General Trend and Required Properties of Well and Drilling.

— Trend —

1. Deep well
2. Low ambient temperature
3. Variety of compositions (Sour, Mixed)
4. Directional drilling
5. Enhanced oil and gas recovery (Water, gas, chemical, fire···)

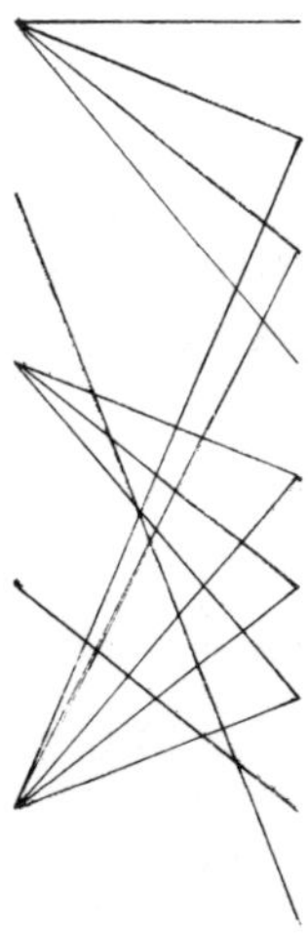

— Requirement —

1. High strength
2. High strength (at elevated temperature)
3. Thermal shock resistance
4. High collapse strength
5. H.S.C.C resistance
6. H.I.C resistance
7. Corrosion resistance
8. Non magnetic properties
9. Low temperature toughness

combinations shown in Figure 4 are considered appropriate with respect to the routes and means of transportation that connect the places of consumption in the United States, Europe and the Far East with the Arctic Circle. The most important subject may

Table 5. Evaluation and Future in Each Type of Transportation System.

Type		Evaluation			Subjects for development	Future
		for Ice	Continuity of supply	Technical completeness		
1. Ship	Ice breaking tanker	◎	○	○	—	Decide by economic evaluation
	Tanker supported by icebreaker	◎	△	○	—	
	Submarine tanker	◎	○	○	—	
	Tanker From transit port	—	◎	◎	Transportation methods for transit port	Combination of those types are suitable choice for future transportation system
2. Pipe line	Sub-sea	◎	◎	◎	Cope with icing	
	On shore	—	◎	◎	Cope with permafrost	
3. Gathering and flow line		—	—	—	Cope with corrosion by untreated oil and gas	

be the development of means of transporting oils or gases produced offshore to nearby shores or transit ports.

Figure 4. Transportation Route for Oil and Gas Produced at Arctic Offshore.

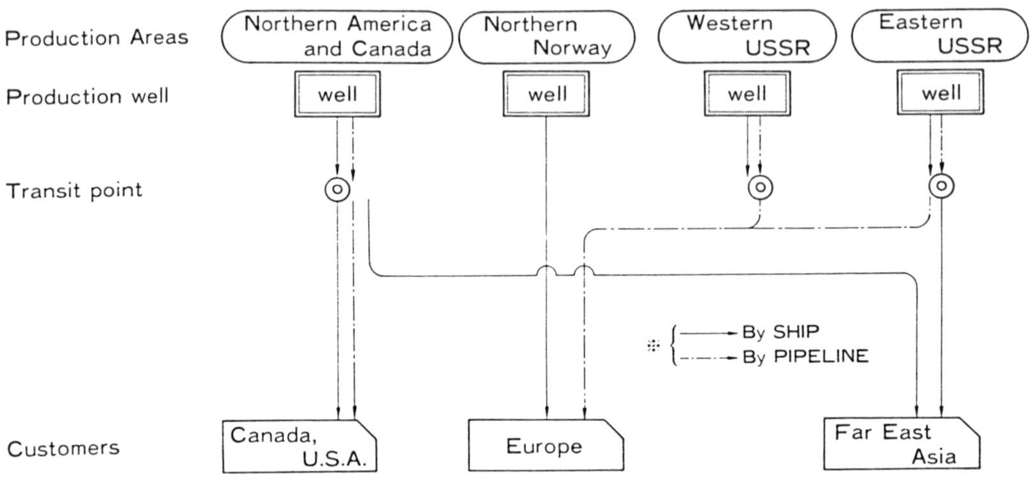

PROPERTIES REQUIRED FOR STEELS

Steels for Offshore Structures and Ships

Table 6 summarizes the properties required for steels. The plate thickness varies depending on the type and part of structures. A maximum plate thickness of 200 mm may be required in production structures. In the case of icebreakers, steel plates with σ_B of 60 kg/mm^2, about 75 mm maximum in thickness, may be used in the bow. For these steels, low-temperature toughness is the most important property. A service temperature of -60°C minimum is presumed. The subject may be what criterion for low-temperature toughness should be adopted. The fracture of structures used for a long period is mostly caused by fatigue. Fatigue fracture is largely controlled by designs and construction procedures that lessen stress concentrations. For this purpose, it is necessary to gather sufficient data on fatigue characteristics in seawater.

The corrosion resistance of structures in seawater is secured by cathodic protection, (using a galvanic anode system and/or impressed current system) and by coating. The hull plate of icebreakers is especially required to provide this corrosion resistance in seawater. It is necessary to ensure well-balanced corrosion resistance of steel plates and welds.

In welding fabrication, there is a trend toward adoption of large-heat-input welding to improve construction efficiency. It is necessary to develop steels, welding materials and welding processes that ensure the toughness of welds under this condition. Furthermore, as a matter of course, steels with low susceptibility to weld-cracking are demanded.

Table 6. Required Properties of Steels for Offshore Structures and Ships.

Items	Drilling and production structures	Transportation ship and supply boats
1. Thickness	~100, ~200mm	~100mm
2. Strength	$\sigma_B 50 \sim 60$, ~80 kg/mm²	$\sigma_B 50$, 60 kg/mm²
3. Toughness	Service temperature ~ −50°C, ~ −60°C	Service temperature −50°C, ~ −60°C
4. Fatigue	Corrosion fatigue property in the sea water	—
5. Corrosion resistance	—	Corrosion resistance for steel without coating for icebreaker
6. Weldability	Toughness for the variety of welding heat input No preheating	Toughness for the variety of welding heat input No preheating

High-Strength, Sour-Resistant Steels

The properties shown in Table 4 are required depending on well depths and environment. What is worthy of special attention in the future is the development of steels that combine high strength and sour-resistant characteristics at high temperatures and steels that are suited to techniques for secondary and tertiary recovery.

Table 7 shows the properties that line pipe is required to provide. It is expected that the sour-resistant characteristics will be strongly required for flow lines and gathering lines. For trunk lines, the requirement for avoiding unstable ductile fracture in the case of rich gas transportation may be added, not to mention brittle fracture toughness.

There are new needs for injection line pipe which transmits CO_2 and water. In this case, properties such as high strength, sour-resistant characteristics and corrosion resistance will be strictly required.

DEVELOPMENT OF STEELS—PRESENT SITUATION AND TRENDS

Steel Plates, Steel Pipe and Line Pipe

For each level of strength and toughness, a system of basic steel plates to be used in various steel structures is expected to be developed after some improvements and modifications. Figure 5 shows the principal reasons for rapid progress in this field over the past several years:

- clarification of needs through close dialogues among steel makers, fabricators and end users related to energy exploitation in the Arctic

- progress in the TMCP (thermo-mechanical control process) and techniques for producing very pure and clean steels.

Table 7. Required Properties of Steel Pipes for Pipelines.

Applications	Requirements for line pipes
1. Trunk line	① High strength ② Unstable ductile fracture resistance for dense gas ③ Low temperature toughness ④ Resistance to hydrogen-induced cracking ⑤ Weldability
2. Flow gathering line	① Low temperature toughness ② Resistance to hydrogen-induced cracking ③ Corrosion resistance (CO_2+H_2S + salt, water, elevated temperature gas and oil) ④ Weldability
3. Injection	① High strength ② Low temperature toughness ③ Resistance to hydrogen-induced cracking ④ Corrosion resistance ⑤ Weldability

Figure 5. Variety of Steel Materials for Offshore Structure Application.

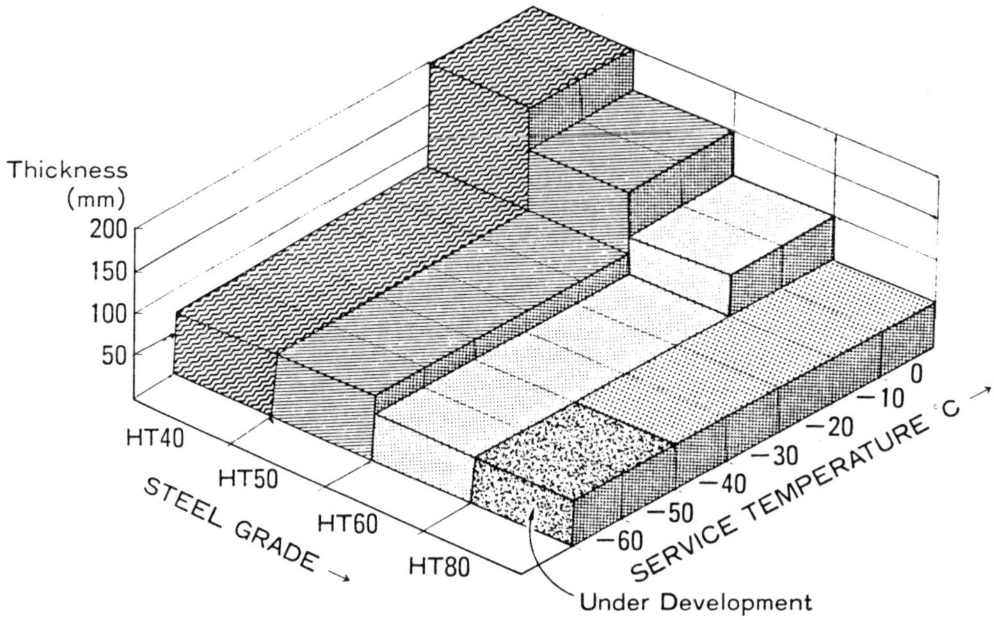

Future principal development issues are:

- steels of various strength and toughness levels for which proper welding processes are considered
- heavy-gage steel plates for low-temperature services
- composite materials with super alloy steels, nonferrous metals and nonmetallic materials.

The greatest development issues of oil country tubular goods are:

- steel pipe that can be used in a high-temperature sour environment
- steel pipe that is used for various kinds of secondary and tertiary recovery.

With respect to the former, the most important characteristic in the high-temperature range is resistance to general corrosion. The relationship between alloying elements, such as chromium and nickel, and the corrosion characteristic in a service environment (composition and temperature) has been grasped; steel pipe of this type is gradually being produced in serial form.

In the latter case, developments that meet individual needs are presently being pushed forward on the basis of the past achievement of each technique. The developments will be smoothly carried out if conditions are made clear.

High-strength steel pipe to X70, sour-resistant steel pipe conforming to the NACE level and low-temperature steel pipe for -50°C minimum have already been used in large quantities as line pipe. Great future issues are:

- development of sour-resistant line pipe (for undehydrated gas and oil without inhibitor)
- development of line pipe for fluid injection.

With respect to the former, it is necessary to develop high alloy steels in view of the characteristics required. In consideration of strength and sour resistance, dual-phase stainless steels have been developed and are being put into practical use. Super alloys will be used in more severe applications. In pursuit of economy, development of manufacturing techniques for clad steel pipe is actively being conducted.

With respect to the latter, excellent sour resistance and corrosion resistance are required of both injection-side and recovery-side line pipes in the case of CO_2 and water injection. Stainless steel pipe, heavily coated pipe, non-metallic pipe, etc., are necessary. In any case, development of composite pipe based on steel will be an issue in the future.

Manufacturing Techniques of Steels

Figure 6 shows the plate manufacturing process as an example of a system of manufacturing techniques of steels already established. This process has the following features:

- techniques for producing large amounts of very pure and clean steels
  —hot metal pretreatment (desulfurization and desphosphorization)
  —secondary refining
  —continuous casting
- development of the means of securing mechanical properties in the rolling process (e.g., TMCP)
  —highly controllable low-temperature reheating furnace
  —heavy-duty rolling mill
  —controlled cooling equipment
- perfect quality and order control by all online control systems.

The basic technical level shown in this example has also been attained in steel pipe (large-diameter pipe mill, seamless pipe mill, electric welded tube mill and hot extrusion press). Therefore, the steel industry can now develop and produce steels that meet various needs if appropriate lead time is given. An issue for the future will be manufacturing techniques of composite materials based on steel. For example, clad steels of dual-phase stainless steel and clad steels with Ni-Cu alloys.

Figure 6. Typical Integrated Plate Production Process.

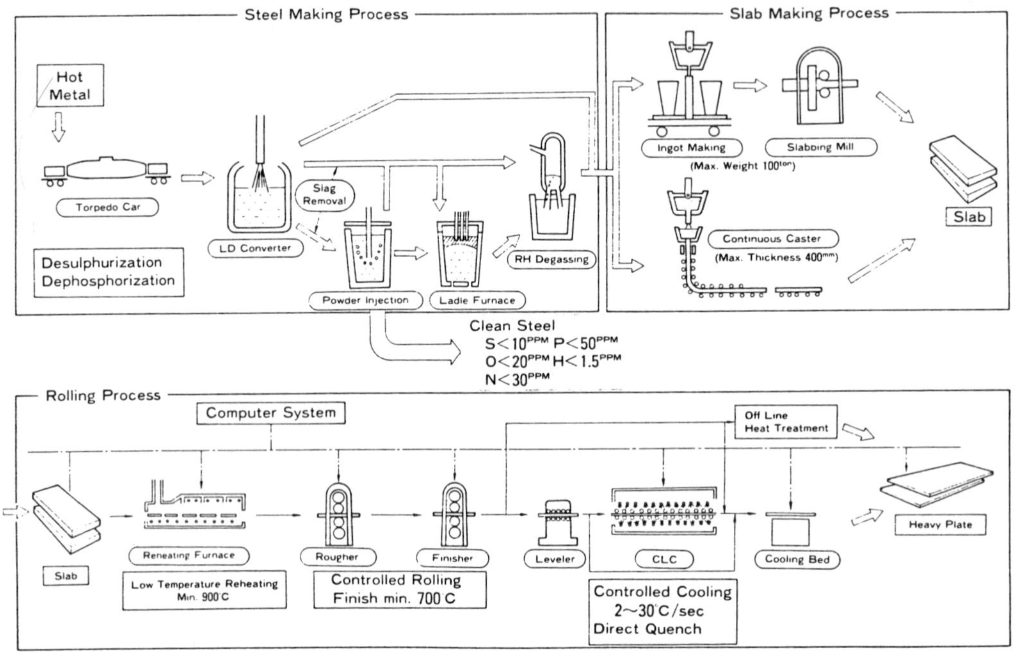

RELATED TECHNIQUES

Steels are goods which are designed so that their performance can be fully displayed under working and service conditions. To produce such steels, it is necessary that organic linkage and cooperation exist between steel metallurgists and specialists of many engineering departments. This section briefly describes future prospects of fracture mechanics, welding and corrosion protection most closely related to steels.

Fracture Mechanics

Table 8 shows future problems in these fields for each mode of fracture of structures. Ductile fracture initiates mostly from stress concentrators. Appropriate designs can now be made, owing to the progress in stress analysis. Most fractures are caused by fatigue fracture. This is due to insufficient stress analysis, unclear design load conditions, some crack-like defects that could not be detected, etc. Therefore, it is also important to establish a basic concept of repairing (including inspection intervals) based on a fail-safe design criterion. It is to meet this practical need that studies on fatigue-

Table 8. Subjects for Fracture Mechanics.

	Fundamental subjects	Practical subjects	
Ductile fracture	——	——	
Fatigue fracture	① Thickness effect on fatigue crack initiation ② Clarification of effect random loading (modification of Miner and Paris equation) ③ Corrosion fatigue mechanism	① Improvement technology of weld toe configuration ② Establishment of standard for repair welding ③ Accumulation of corrosion fatigue data (initiation and propagation)	For design Collection of reliable data for design conditions and precise evaluation methods
Brittle fracture	① Significance of pop-in in COD test ② Brittle fracture strength under dynamic loading	① Establishment of reasonable required COD values by considering the importance of structural members	

crack propagation rates are actively conducted. It seems that principal basic problems are the following:

- clarification of the effect of random loads and establishment of design specifications in which results are reflected
- accumulation of data on corrosion fatigue and clarification of fundamental mechanism.

Brittle fracture is considered to initiate from fatigue cracks in structures that have many stress concentrators and are subjected to fluctuating loads such as offshore structures. There are stress-intensity factor K, J-integral and COD as concepts of brittle fracture initiation. From practical viewpoints, however, COD is judged to be appropriate. The concept of COD is adopted in design codes such as British Standard and W.E.S. (Japan). Therefore, practical future subjects are:

- clarification of the significance of pop-in type fracture
- clarification of required fracture toughness for each position of structural members.

Furthermore, it is necessary that results should be reflected in design codes.

Brittle-fracture arrest characteristics are not required in jacket-type structures where uniform stress fields (as in ships) do not exist but instead, structural redundancy exists. Incidentally, the clarification of the brittle-fracture behavior, under dynamic loads, upon collision with floating ice, is one of the basic issues of the future.

Structures cannot be produced economically if all pre-existing defects and service conditions of structures are supposed to be the severest ones in all fractures. It seems to be important to have an accurate grasp of data on these basic design conditions and to treat them properly.

Welding Processes

A reliable and cost-effective joining method for the future, suitable for steel structures, will be the fusion welding process. Four development issues related to the welding of steel structures to be used in the Arctic are:

1. development of welding processes, welding materials and steels that are appropriate in terms of weld-crack susceptibility, fracture toughness and efficiency to cope with increased strength and plate thicknesses
2. development of welding processes that are suitable for the remodeling of steel structures used for a long time or the development of welding processes needed in case of unexpected trouble
3. reliable automated processes in 1 and 2 above
4. application of new heat sources to welding.

With respect to 1 above, what is called large-heat-input welding processes (including welding materials and steels) have been developed and put into practical use, suitable for presently used plate thicknesses and required fracture toughness. When future trends are taken into consideration, however, the narrow gap welding process is considered to be appropriate in terms of fracture toughness, welding efficiency and, above all, cost. Many types of this welding process have been developed and put into practical use as, for example, with welding reactor pressure vessels. (See Table 9.) The lack of penetration, which has so far been the defect of this welding process, was conquered by using present advanced electronics engineering. Moreover, it is predicted that full automation of the whole process of multi-pass welding will be realized. At present, fully automatic robot welders are in use in the shipbuilding industry to produce fillet welds, which are large in number for welded joints. In the future, they will be applied in further advanced forms to structures to be used in the Arctic.

With reference to 2 above, various wet or dry-type welding processes are presently used at small water depths. In the future, however, the necessity for welding at larger depths will increase and it is expected that completely dry-type automatic welding processes by remote control will be developed.

Concentrated high-density energy includes electron beams, laser and plasma. These are used in various industrial fields for heating, cutting, joining, heat treatment, etc., by making the most of their features. Electron beams are used for joining pressure vessels, shafts and gears; lasers are used for joining materials of small thickness and small parts. Both are expected to develop into ultimate narrow gap welding processes in the future. For this purpose, the following problems must be solved:

- development of steels suitable for those two welding processes
- development of methods of controlling the toughness of welds
- development of higher-output lasers and methods of beam focusing.

Corrosion Protection

The only defect of steel is corrosion—the last challenge to steel manufacturers. Complete corrosion protection is required from the standpoint of corrosion fatigue prevention in addition to the prevention of local or general corrosion. Offshore structures in the Arctic Circle may require corrosion protection techniques of a higher level than in other districts because of the following conditions:

Table 9. Variations of 'SUBNAP'' Process and its Features.

Method	Edge preparation and pass sequence	Applicable plate thickness mm	Features	Points to be considered
The small angular-X-groove method	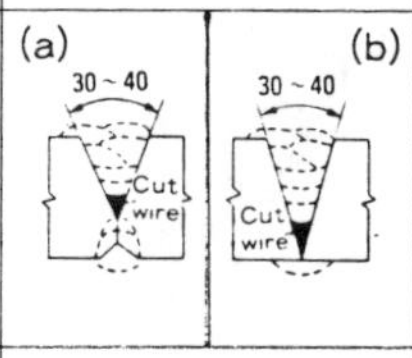	~80	○ Most effective in saving of welding time and materials for the plate thickness 40 to 80 mm ○ Capable of no back-gauging, method(b)	○ Unfused part in root needs to be removed on back gauging or welding of the 1st pass of backside. ○ Use of special flux of excellent slag removal
The single center pass method	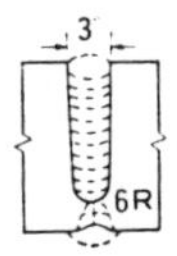	~150	○ Most effective in saving of welding time and materials for the plate thickness over about 70 mm	○ Hot crack on 1st pass on high C steel ○ Fairly severe control of welding conditions ○ Use of special flux of excellent slag removal
The two-pass layer method	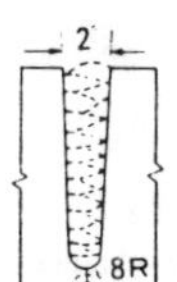	~300	○ Most applicable to over 150 mm thick plate ○ Easy slag removal ○ Wide range of allowable welding conditions ○ Excellent toughness by small heat input	○ Use of special flux of good slag removal and good bead contour

- long service periods of 20 to 40 years
- damage by ice
- environment in which repairs for corrosion protection are difficult.

Cathodic protection with sacrificial anodes, impressed currents and protective coating and combinations of these types of corrosion protection are presently applied to offshore structures.

The severest corrosion in seawater occurs in the splash zone. The most positive method of corrosion protection available at present is the coating with Ni-Cu alloys or Monel alloys. The following are expected to progress in this field in the future:

- new corrosion protective coatings, including those for repairing (e.g., flexible or hard coatings, such as urethane, polyethylene—with characteristics suitable for ice—and new materials, such as ceramics
- defect detection systems
- repair techniques.

Urethane coatings have already been put into practical use. If conditions are made clear when steel structures are in contact with ice, it is possible to develop materials with characteristics necessary for the conditions. Further development of this system is expected. (See Figure 7.)

CONCLUSION

The present author is associated with the material makers who must develop and supply appropriate goods to meed the needs of users. This symposium is concerned

Figure 7. Types of Protective Coating.

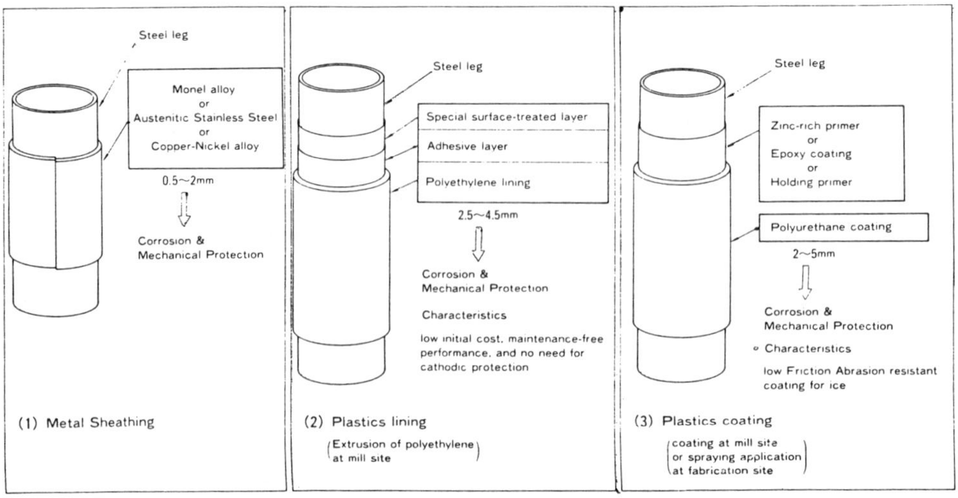

with the future needs that steel manufacturers will be required to meet. He hopes that trends in energy exploitation in the Arctic Circle and the present situation of makers of other materials competing with steels, will be well understood and that new development themes will be given to steel manufacturers. He also hopes that, due to the momentum of the symposium, a closer exchange of information will be made concerning more concrete needs, improving steels to meet those needs and techniques for manufacturing such steels.